Research, Innovation and Development in Engineering (Expotecnologia 2021)

IOP Conference Series: Materials Science and Engineering
Volume 1253

Online
3 - 5 November 2021

ISBN: 978-1-7138-6153-9
ISSN: 1757-8981

Printed from e-media with permission by:

Curran Associates, Inc.
57 Morehouse Lane
Red Hook, NY 12571

Some format issues inherent in the e-media version may also appear in this print version.

Institute of Physics
Dirac House, Temple Back
Bristol BS1 6BE UK

Phone: 44 1 17 929 7481
Fax: 44 1 17 920 0979

techtracking@iop.org

Additional copies of this publication are available from:

Curran Associates, Inc.
57 Morehouse Lane
Red Hook, NY 12571 USA
Phone: 845-758-0400
Fax: 845-758-2633
Email: curran@proceedings.com
Web: www.proceedings.com

TABLE OF CONTENTS

Expotecnología 2021

Organized by

Sponsor

IOP Conf. Series: Materials Science and Engineering 1253 (2022) 011001 doi:10.1088/1757-899X/1253/1/011001

Expotecnología 2021 was an event that took place in the city of Cartagena de Indias, from November 3 to 5, 2021. With the slogan *"Engineering and Research Practices for Sustainable Development"*, this nineteenth version was a meeting that brought together the scientific and academic community in spaces for the dissemination and transfer of knowledge, where innovation was the essence and a fundamental part of the event. Expotecnología was created in the Faculty of Engineering Sciences - FACI, of the Antonio de Arevalo University Foundation - Unitecnar in collaboration whit Antonio Jose de Sucre University Corporation - Corposucre in this version it maintained as mission, the development of the interaction between science, technology and society, through establishing of knowledge networks and academic and professional cooperation. The main objective was to address the challenges that emerge from the everyday life of the engineering community, the problems immersed in the development of projects and programs of Higher Education Institutions (IES), organizations and research centers at the national and international level. In the development of the day of presentations in a virtual way, it was possible to promote the dissemination of the research work of professionals, teachers and students from all areas of engineering, framed in four (4) thematic axes, as follows:

International Congress of New Technologies

- Big data, artificial intelligence
- Business intelligence
- Software Engineering
- Education and new technologies
- Mobile apps.

International Congress of Petrochemical and Process Control

- Material sciences
- Instrumentation and control systems
- Robotics and automation
- Industrial maintenance

International Congress of Electricity, Electronics and Telecommunications

- IoT, fiber optics
- Digital signal processing
- Wireless networks
- Mobile communications
- Power systems
- Energy storage
- Smart grids
- Renewable energy
- Energy markets.

Construction Summit

- Modern building systems

• Computational modeling.

On this occasion, the event was attended by 450 people, representatives from 15 national and **12** international university. 97 virtual presentations were made, of which 34 articles were selected for this volume of the IOP Conference Cycle: Materials Science and Engineering and 19 articles for the book Engineering, Innovation and the fourth industrial revolution. After a process of Peer Review. The plenary conferences were in charge of 2 national and 8 international speakers.

Invited speakers

Dr. Leandro Lodi Chemical Engineering Professor Federal University of Alfenas - UNIFAL-MG Campus Poços de Caldas MG - Brasil

Dr. Aryanne Quintal Project Coordination Section of Competitiveness, Innovation And Technology Organization of American States (OEA)

Msc David Chang Villacreses Duke Office of translation and commercialization (OTC)

Dr. Luis Felipe Ariza Vesga Director of Research and Development ID:TOLÚ

MSc. Ing. Guido Caicedo Rossi Executive Director of I3LAB Professor of Engineering at ESPOL

Ing Eduardo Santos Ladino Senior Offshore Defense Advisor Ministry of Defence General Maritime Directorate Republic of Colombia

PhD. Jose Luis Guerrero Cusumano Associate Professor - Georgetown University

Ing Jesus Antonio Garcia Sanchez Professor Mechanical Engineering Itajuba University

Ing Roberto Giordano Lorena Dean Faculty of Engineering FASTA University

Mabel Gisela Torres Ex-Minister science technology and innovation Republic of Colombia

IOP Conf. Series: Materials Science and Engineering 1253 (2022) 011001 doi:10.1088/1757-899X/1253/1/011001

University Participants

Organizing Committee

President
Msc. Libis Valdez Cervantes, Antonio de Arevalo University Foundation

Comité Editorial
Msc. Libis Valdez Cervantes, Antonio de Arevalo University Foundation,
Msc. Yesid Tarriba Lezama, Antonio de Arevalo University Foundation
Msc Sergio A Sanchez Hernandez, Antonio Jose de Sucre University Corporation
Msc Alex Davis Morales Acosta, Antonio Jose de Sucre University Corporation

Comité técnico
Msc. Yesid Tarriba Lezama, Unitecnar
Msc. Libis Valdez Cervantes, Unitecnar
Msc Sergio A Sanchez Hernandez, Corposucre
Msc Alex Davis Morales Acosta, Corposucre

IOP Conf. Series: Materials Science and Engineering 1253 (2022) 011001 doi:10.1088/1757-899X/1253/1/011001

Scientific Committee
Msc. Libis Valdez Cervantes, Unitecnar, President (Colombia)
Msc. Yesid Tarriba Lezama, Unitecnar, Coordinator (Colombia)
Msc Alexander Carabali Alvarado (Colombia)
Phd Arly Dario Rincon Quintero (Colombia)
Msc Brayan Tarazona Romero (Colombia)
Msc David Javier Muñoz Aldana (Colombia)
Msc Eduardo Duque Cuesta (Colombia)
Msc José Ruiz-Meza (Colombia)
Phd John Castro Suarez (Colombia)
Msc Nancy Tavera Castillo (Colombia)
Msc Sergio Antonio Sanchez Hernandez (Colombia)

Special thanks

We greatly thank all the authors for their valuable participation in Expotecnología 2021 and all the peer reviewers for their commitment and dedication in reviewing the papers to be published in IOP and Book chapters.

Dr. Libis Valdez Cervantes
Msc. Yesid Tarriba Lezama
Editors of the Proceedings of EXPOTECNOLOGIA 2021

IOP Conf. Series: Materials Science and Engineering 1253 (2022) 011002 doi:10.1088/1757-899X/1253/1/011002

Peer Review Statement

All papers published in this volume have been reviewed through processes administered by the Editors. Reviews were conducted by expert referees to the professional and scientific standards expected of a proceedings journal published by IOP Publishing Publishing.

- **Type of peer review:** Double Anonymous
- **Conference submission management system:** Morressier
- **Number of submissions received:** 29
- **Number of submissions sent for review:** 24
- **Number of submissions accepted:** 24
- **Acceptance Rate (Submissions Accepted / Submissions Received × 100):** 82.8
- **Average number of reviews per paper:** 2
- **Total number of reviewers involved:** 12
- **Contact person for queries:**
  Name: Yesid Tarriba Lezama
  Email: expotecnologia@unitecnar.edu.co
  Affiliation: UNITECNAR - Faculty of Engineering

Study on the gasification process of the biomass obtained from agricultural waste with the purpose of estimating the energy potential in the Santander region and its surroundings.

M A Durán-Sarmiento [1,2], L A Del Portillo-Valdés[2], Y J Rueda-Ordonez[3], J S Florez-Rivera[1] and A D Rincón-Quintero [1,2]

[1] Design and materials research group (DIMAT), Automation and Control Energy Systems Research Group (GISEAC), Electromechanical engineering, Faculty of Natural Sciences and Engineering, Technological Units of Santander (UTS), Calle de los Estudiantes No 9-82, 680005, Bucaramanga, Colombia.
[2] Energy in Building research group (ENEDI), Doctoral Program in Energy Efficiency and Sustainability in Engineering and Architecture, Department of Machines and Thermal Engines, University of the Basque Country, UPV/EHU, Engineer Torres Quevedo Plaza, 1, 48013, Bilbao, Spain.
[3] Energy and environment research group (GIEMA) Department of Mechanical Engineering, Industrial University of Santander, UIS, 680002, Santander, Colombia.

aduran@correo.uts.edu.co

Abstract. The generation of energy through renewable sources motivates to study the process of gasification of agricultural residual biomass and to obtain the greatest benefit. In turn, the energy potential in the department of Santander was estimated, which represents 2% of the total production nationwide, and an average of the energy production that can be obtained with the gasification process was obtained, yielding a result of 67% efficiency, with this the department of Santander is located in the ranking of bioenergy producing departments in Colombia. As an additional factor, an economic analysis was carried out to calculate an average of how much money can be saved by replacing fossil fuels with biofuels. The gasification of residual biomass is one of the most striking and studied methods today, with which you can contribute positively to the environment and with which you can achieve a bioeconomy quite sustainable and independent of fossil fuels.

1. Introduction.

Biomass as a clean and sustainable energy source of carbon has been used since the beginning of time until approximately the industrial revolution, these needs range from the production of metals to power steam engines, these new and current utilities decreased the use of biomass [1]. With the passage of time and year after year the residual biomass has regained more and more strength as a clear futuristic option with which you can reduce the use and consumption of fossil fuels to meet energy needs, in addition to this you have the possibility of contributing in a collateral way to the improvement of the global environment, knowing that currently this purpose is a worldwide trend.

Energy consumption in recent years has increased due to the expansion of industrial processes and urban areas, this increase has produced interest in reducing the use of fuels derived from petroleum, in order to promote the consumption of alternative and renewable energy sources. [2].

For this reason, it has been chosen to carry out an investigation on the benefits and the use obtained from the residual biomass specifically of the agricultural residual biomass, this research has a focus on the department of Santander that annually generates millions of tons of residual biomass in the agricultural, livestock, and urban sector, the documentary base taken as a reference is the Atlas of the Energy Potential of Residual Biomass in Colombia [3].

2. Work development

According to Schallenberg "energy refers to the ability that bodies can have to originate work, such as heat generation, light emission, etc." Most of the energy that the planet has is of solar origin, their source of light and heat are the main elements by which they develop different chemical reactions and thanks to these reactions help living beings to evolve, being the main material for the production of energy [4], energy can be classified into conventional energy, which refers to that which does not consist of a high level of evolution in technology to be able to obtain energy from the source in question and unconventional energy that refers to energy that requires a high level of development in order to achieve its use [5].

When we talk about biomass, we mention all the living matter that may be exposed on the surface of the soil to which it is called the biosphere. Plants are the ones that initiate this process through photosynthesis capturing the energy of the sun and biomass as it belongs to the repetitive circle of natural cycles is considered to be renewable [3], the energy value of biomass depends on the proportion of energy it has and that it could release at the time it is subjected to an energy transformation, so it is considered that biomass is a very attractive source to be used energetically, according to British Petroleum indicates that Colombia in 2017, presented an increase in the consumption of renewable energies of 7.9%, this indicates that the importance of making more use of this energy resource has begun to be noticed in a country that has a continental area of 114,174,800 hectares [6].

In the industrial and agricultural environment of the country there are several sources of residual biomass, such as the agricultural sector, livestock and organic solid waste or urban waste. In order to carry out the research, it focused only on agricultural residual biomass since it is a quite attractive option for energy generation, then in Figure 1 the different biomass sources are mentioned.

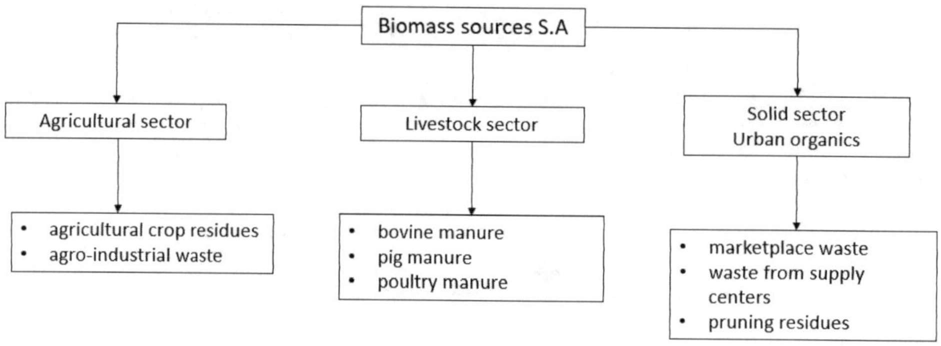

Figure 1 Biomass sources.

According to the classification presented in figure number 1, an analysis of the energy potential contained in agricultural residual biomass in Colombia is carried out, taking into account the different crops and departments. Next, table 1 presents the consolidated energy potential of agricultural residual biomass per crop in Colombia, which contains a total production of 14,974,807 tons per year, and a total energy potential per crop of 331,645.68 tera joules per year.[3].

Table 1. Energy potential of agricultural residual biomass by crops in Colombia.

Cultivation	Energy Potential TJ/year
Oil Palm	16.013,64
Sugar cane	118.578,87
Cane Panela	81.054,57
Coffee	49.106,88
Corn	20.802,79
Rice	27.835,94
Banana	6.595,92
Plantain	11.657,07
Total	331.645,68

Below, Table 2 presents the total energy potential of agricultural residual biomass in Colombia detailed by departments that contains a total sown area of 3,003,064 hectares with a total production of 71,943,162 tons per year, this for the production of a total energy potential of 11,657.07 tera joules per year [3].

Table 2. Energy potential of agricultural residual biomass by department in Colombia.

Department	Energy potential [TJ/year]	Department	Energy potential [TJ/year]	Department	Energy potential [TJ/year]
Antioquia	2,063.92	Huila	409.43	V. Cauca	516.5
Atlántico	6.09	La Guajira	48.76	Arauca	258.66
Bolívar	114.35	Magdalena	52.81	Casanare	112.26
Boyacá	121.06	Meta	859.52	Putumayo	148.02
Caldas	641.65	Nariño	501.72	Amazonas	7.05
Caquetá	284.83	Norte de Santander	258.85	Guainía	2.9
Cauca	457.41	Quindío	1,161.94	Guaviare	74.6
Cesar	72.69	Risaralda	683.55	Vaupés	5.52
Córdoba	970.16	Santander	206.13	Vichada	3.83
Cundinamarca	261.75	Sucre	26.83	TOTAL	11,657.07
Choco	359.11	Tolima	965.15		

In order to meet the general objective, a study is carried out on the percentage of the amount of synthesis gas that can be obtained through the process of gasification of agricultural residual biomass with some efficiency of the process. To reach this conclusion, research is carried out in a certain number of articles, books and scientific journals where sufficient information is found to be able to obtain an average efficiency value for the gasification process and what can be taken advantage of from agricultural residual biomass, in order to replace the energy derived from fossil fuels with bioenergy, see table 3.

Table 3. Percentage of use by means of gasification.

	Reference	**Theme**	**Title**	**Approximate gasification efficiency**
1	[7]	In this study, the analysis and comparison of the simulation results of a gasification system is carried out.	Potential power generation on a small-scale separated-type biomass gasification system. [7].	73,8%
2	[8]	In this work CLG is applied to microalgae to obtain synthesis gas through a fixed bed reactor.	Characteristics of microalgae gasification through chemical looping in the presence of steam. [8].	61,65%
3	[9]	In this work, a model was developed which allows to simulate a biomass gasification process for the production of synthesis gas.	Wood gasification to obtain a syngas useful in the production of biofuels and / or chemical products. [9].	72%
4	[10]	In this study the simulation of the gasification process is carried out for jatropha seed shells and stems of Dichrostachys cinerea.	Simulation of the biomass gasification process from Jatropha urcas and Dichrostachys cinérea. [10]	70%
5	[11]	In this experimental investigation, rice straw was used as biomass to gasify it in a two-stage fluidized bed reactor.	Experimental study of biomass gasification with oxygen-enriched air in fluidized bed gasifier. [11]	33,78%
6	[12]	Bio-oil derived from the pyrolysis of rice husk is gasified to produce synthesis gas, and the effectiveness of the gasifying agents was analyzed.	Gasification of bio-oil: Effects of equivalence ratio and gasifying agents on product distribution and gasification efficiency. [12]	75%
7	[13]	In this article two schemes were proposed to increase the gasification efficiency of the depolymerization slag.	Enhancement of depolymerization slag gasification in supercritical water and its gasification performance in fluidized bed reactor. [13]	71%
8	[14]	In this work, the conversion of dried sewage sludge was carried out by means of pyrolysis and gasification processes in order to obtain synthesis gas.	Production of Hydrogen-Rich Syngas from Gasification and Pyrolysis of Dry Solar Wastewater Sludge: Experimental and Modeling Investigations. [14]	82%
9	[15]	A newly developed gasification process utilises waste plastics, carbon-containing sorting residues and rubber parts as well as shredded materials in the automotive industry.	Generating syngas from plastic wastes. [15]	80%
10	[16]	As the demand for energy increases, the need to take advantage of renewable energy sources increases.	Bioenergy from phytoremediate phytomass from aquatic plants via gasification. [16].	50%

According to the table described above which contains the information obtained from different articles and authors who carried out on-site tests and theoretical tests of the percentage of use that can be obtained in the process of biomass gasification. From these data, an average value of 67% was determined, this percentage is the average theoretical value that was obtained from the research that was carried out and is the percentage that can be used at the time of applying the gasification process to agricultural residual biomass in Colombia.

In the process of gasification of agricultural residual biomass, the equivalence ratio (ER) is an important factor, this aspect is the relationship between oxygen or supply air and oxygen or air necessary to obtain a complete and efficient combustion, also for the design of a gasifier this component is important because it helps to sustain the quality levels of the final product that in this case is syngas or syngas. In the present study, the analysis of the process of gasification of agricultural residual biomass is carried out with equivalence ratios (ER) ranging from 0.1 to 0.5, with gasifying agents such as air, air enriched with oxygen and pure oxygen [17]. See table 4.

Table 4. Efficiency of gasifying agents.

ER	**Air + Oxygen Efficiency**	**Air**	**Oxyge n**
0,10	53,59	55,84	54,58
0,15	58,35	59,67	58,29
0,20	63,78	64,37	62,78
0,25	71,38	70,72	70,72
0,30	76,21	74,36	75,48
0,35	65,76	65,76	65,76
0,40	51,27	53,46	52,40
0,45	43,60	46,05	44,79
0,50	38,77	40,36	39,50

3. Results and discussions.

With the research carried out, a theoretical average value of 67% efficiency for the gasification process is determined. According to the data presented where the equivalence ratio for gasifying agents air, air enriched with oxygen and pure oxygen is exposed, it can be evidenced that the best ER equivalence ratio for the crop gasification process is 0.3 and it can also be concluded that the best gasifying agent is pure air.

If the gasification process is used and applied to the entire energy potential of agricultural residual biomass by crops in Colombia, a total energy potential of 222,202.61 tera joules per year can be obtained, in the same way if the process of gasification of agricultural residual biomass by departments in Colombia is applied, a total energy potential of 7,810.22 tera joules per year can be obtained.

Making a comparison of the department of Santander with respect to the department of Antioquia being the department with the most production of energy potential from biomass in Colombia, this in order to observe the energy potential present in Santander and the energy potential that can be used as stated by different authors which describe different methods and efficiencies for the process of gasification of agricultural residual biomass.

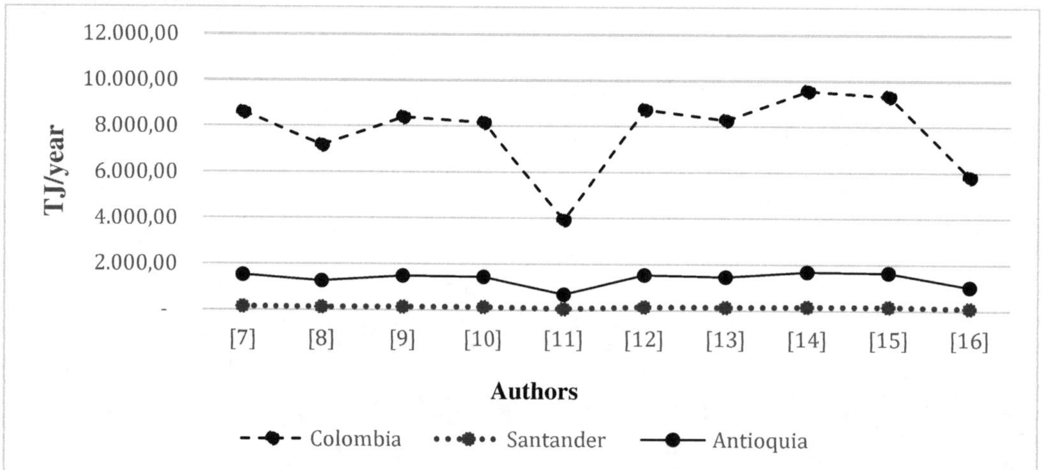

Figure 2. Comparison of energy potential through gasification of agricultural residual biomass

If the gasification method for agricultural residual biomass is used and implemented, a certain amount of money can be saved, due to the replacement of energy from fossil fuels with bioenergy. To justify the above written, an example is presented with the department of Santander. According to the mining-energy planning unit (UPME) for the year 2017 Colombia generated a total of energy of 66,666,991 gigawatts per hour and the department of Santander participated in the generation of energy with 5,632.55 giga watts per hour that is equivalent to 8% of the country's energy share.

In Colombia, a total of energy is generated annually from residual biomass of 11,657.07 tera joules per year and in the department of Santander an amount of energy is generated annually from residual biomass of 206.13 tera joules per year, if the comparison of the department of Santander with respect to Colombia is made, the department of Santander generates 2% of energy from residual biomass. The energy produced by the residual biomass in Santander is 57,260,000 kilo watt per hour if with this energy the gasification process is used, the result is 38,364,200 kilo watt per hour, then this would be the energy that can be generated in Santander from the use of the gasification process of the residual biomass. Table

5 and 6 presents the approximate amount of money that can be generated if the gasification process is carried out for agricultural residual biomass in Colombia.

Table 5. Residential rates.

Economic Representation		
Stratum	**Residential Rates [KWh]**	**Money Generated Annually**
1	$ 238,76	$ 9.159.836.392
2	$ 298,46	$ 11.450.179.132
3	$ 507,38	$ 19.465.227.796
4	$ 596,92	$ 22.900.358.264
5	$ 716,30	$ 27.480.276.460
6	$ 716,30	$ 27.480.276.460

Table 6. Commercial rates.

Economic Representation		
Measured Level	**Residential Rates [KWh]**	**Money Generated Annually**
I	$ 647,84	$ 24.853.863.328
II	$ 589,36	$ 22.610.324.912
III	$ 485,21	$ 18.614.693.482
IV	$ 446,44	$ 17.127.313.448

A superficial economic analysis is carried out of what the production of biogas can cost through the gasification of the residual biomass. In order to justify the cost of producing biogas in a plant, the research project carried out by H. Thunman, in which they evaluate three important factors for the production of this fuel, first investment cost, second cost of the development of the plant and finally the operating cost.

These costs are evaluated in a simulation of three plants with an efficiency of 70% and with different production capacities, 20 MW (megawatts), 100 MW (megawatts) and 200 MW (megawatts), where the plant with a capacity of 20 MW (megawatts) has the highest production cost and the plant with a capacity of 200 MW (megawatts) presents the lowest production cost H. Thunman. Next, table 7 presents the comparison of the production costs of the three commercial plants, where values in Colombian pesos are presented and evaluated by KWh (kilo watts hour).

According to the analysis carried out, it can be evidenced that the number 1 commercial plant with a production capacity of 20 MW is not economically viable for residential tariffs with strata 1, 2 and 3, and for commercial tariffs with levels III and IV. Similarly, it is evident that the number 2 commercial plant with a production capacity of 100 MW is not economically viable for residential tariffs with strata 1 and 2, and in commercial tariffs it is economically viable for all levels. Finally, it is identified that the number 3 commercial plant with a production capacity of 200 MW is economically viable for all commercial and residential tariffs with the exception of stratum number 1.

IOP Conf. Series: Materials Science and Engineering 1253 (2022) 012001 doi:10.1088/1757-899X/1253/1/012001

Table 7. Production cost for biogas.

Costs Evaluated	Production Capacity		
	Commercial Plant 1: 20 MW, COP/KWh	Commercial Plant 2: 100 MW, COP/KWh	Commercial Plant 3: 200 MW, COP/KWh
Cost of capital, depreciation	188	87	63
Cost of capital, interest (5%)	113	52	38
Cost of development	19	9	7
Operating costs (excluding raw materials)	154	72	58
Cost of raw material	95	95	95
Total cost	$ 569	$ 315	$ 261

4. Conclusions and recommendations.

All the benefits obtained if energy is generated from renewable energy sources and taking advantage of the gasification process are clearly observed, to the point of being able to satisfy the needs of energy consumption, thus achieving independence from fossil fuels that are currently the ones with the highest demand worldwide.

Worldwide one of the priorities is the production of energy derived from renewable sources, based on the benefits it brings to the environment are quite favorable and because of these new green technologies are driving every day in the improvement of their processes to obtain a better performance and use.

When you contribute a percentage of energy generated from renewable sources, this participation, no matter how small it is, will produce economic savings remarkable enough to be able to realize the benefit of generating energy using new green technologies, this can be achieved by applying the different generation methods such as biomass gasification.

Biomass gasification is one of the most studied methods today and one of the most attractive in the future with which you can contribute to make a bioeconomy sustainable enough. These green technologies and their new and innovative processes to produce energy have evolved over time and are gaining strength in all parts of the world.

According to the research carried out where the energy potential from the process of gasification of agricultural residual biomass is presented, it can be clearly seen the benefit it would bring in the economic and environmental field both for the department of Santander and for the entire country.

The university community is invited to continue carrying out research on these green technologies where it can be improved and contribute to the evolution of the processes that lead us towards the economic and environmental future, this research is open to those who may be interested to continue with the study.

5. References.

[1] J. Li *et al.*, "Biomass gasification in molten salt for syngas production," *Energy*, vol. 210, p. 118563, 2020, doi: 10.1016/j.energy.2020.118563.

[2] D. Gielen, F. Boshell, D. Saygin, M. D. Bazilian, N. Wagner, and R. Gorini, "The role of renewable energy in the global energy transformation," *Energy Strateg. Rev.*, vol. 24, no. June 2018, pp. 38–50, 2019, doi: 10.1016/j.esr.2019.01.006.

[3] M. Scharfstein and Gaurf, *Atlas De Biomasa*, vol. 53, no. 9. 2013.

[4] B. Kahraman, "Energy potential of agricultural waste biomass in colombia," 2016.

[5] Natalia Enríquez Dutrús, Use of non-conventional energy. Natalia Enríquez Dutrús," 2018.

[6] British Petroleum Report, "67 th edition Contents is one of the most widely respected," *Stat. Rev. World Energy*, p. 40, 2018, [Online]. Available: https://www.bp.com/content/dam/bp/business-sites/en/global/corporate/pdfs/energy-economics/statistical-review/bp-stats-review-2018-full-report.pdf.

[7] Y. A. Situmorang, Z. Zhao, A. Yoshida, Y. Kasai, A. Abudula, and G. Guan, "Potential power generation on a small-scale separated-type biomass gasification system," *Energy*, vol. 179, pp. 19–29, 2019, doi: 10.1016/j.energy.2019.04.163.

[8] G. Liu, Y. Liao, Y. Wu, X. Ma, and L. Chen, "Characteristics of microalgae gasification through chemical looping in the presence of steam," *Int. J. Hydrogen Energy*, vol. 42, no. 36, pp. 22730–22742, 2017, doi: 10.1016/j.ijhydene.2017.07.173.

[9] L. Suárez-Hernández, J. F. Pérez Bayer, and R. Barrera Zapata, "Gasification of wood to obtain a syngas useful in the production of biofuels and/or chemical products." *Rev. ION*, vol. 30, no. 1, pp. 57–71, 2017, doi: 10.18273/revion.v30n1-2017005.

[10] I. Tobío Pérez, Y. Díaz Domínguez, M. Pfeil, D. Denfeld, R. Piloto Rodríguez, and S. Pohl, "Simulation of the biomass gasification process from Jatropha curcas and Dichrostachys cinerea" *Afinidad*, 2020, [Online]. Available: https://www.raco.cat/index.php/afinidad/article/view/377424.

[11] L. Liu *et al.*, "Experimental study of biomass gasification with oxygen-enriched air in fluidized bed gasifier," *Sci. Total Environ.*, vol. 626, pp. 423–433, 2018, doi: 10.1016/j.scitotenv.2018.01.016.

[12] J. L. Zheng, M. Q. Zhu, J. L. Wen, and R. cang Sun, "Gasification of bio-oil: Effects of equivalence ratio and gasifying agents on product distribution and gasification efficiency," *Bioresour. Technol.*, vol. 211, pp. 164–172, 2016, doi: 10.1016/j.biortech.2016.03.088.

[13] C. Wang, C. Zhu, J. Huang, L. Li, and H. Jin, "Enhancement of depolymerization slag gasification in supercritical water and its gasification performance in fluidized bed reactor," *Renew. Energy*, vol. 168, pp. 829–837, 2021, doi: 10.1016/j.renene.2020.12.104.

[14] A. B. H. Trabelsi *et al.*, "Hydrogen-Rich Syngas Production from Gasification and Pyrolysis of Solar Dried Sewage Sludge: Experimental and Modeling Investigations," *Biomed Res. Int.*, vol. 2017, 2017, doi: 10.1155/2017/7831470.

[15] G. Hirn, "Generating syngas from plastic wastes," *Fed. Minist. Econ. Aff. Energy*, pp. 1–4, 2016, [Online]. Available: http://www.bine.info/en/topics/renewable-energy-sources/biomass/publikation/synthesegas-aus-kunststoffabfaellen-erzeugen/.

[16] S. Kathi, *Bioenergy from phytoremediated phytomass of aquatic plants via gasification*. Elsevier Inc., 2016.

[17] L. Han, M. Zheng, and C. Wang, "Current status and development of terminal blend tyre rubber modified asphalt," *Constr. Build. Mater.*, vol. 128, pp. 399–409, 2016, doi: https://doi.org/10.1016/j.conbuildmat.2016.10.080.

[18] H. Thunman, C. Gustavsson, A. Larsson, I. Gunnarsson, and F. Tengberg, "Economic assessment of advanced biofuel production via gasification using cost data from the GoBiGas plant," *Energy Sci. Eng.*, vol. 7, no. 1, pp. 217–229, 2019, doi: 10.1002/ese3.271.

Evaluation of solar tracking systems applied to small-scale photovoltaic systems: a review

B E Tarazona-Romero[1 2], J S Carreño-Rodriguez [1], A D Quintero-Ruiz[1 2], C L Sandoval-Rodriguez[1 3] and O Lenguerke-Péreza [1]

[1] Automation and Control Energy Systems Research Group (GISEAC), Faculty of Natural Sciences and Engineering, Electromechanical engineering, Unidades Tecnologicas de Santander (UTS), Student Street No 9-82, 680005, Bucaramanga, Colombia.
[2] Energy in Building research group (ENEDI), Doctoral Program in Energy Efficiency and Sustainability in Engineering and Architecture, Department of Machines and Thermal Engines, University of the Basque Country (UPV / EHU), Engineer Torres Quevedo Square, 1, 48013, Bilbao, BI, Spain.
[3] Doctoral Program in Electronic and Communications, Department of Communications Engineering, University of the Basque Country, UPV/EHU, 48013 Bilbao, Spain.

Abstract. This work evaluates solar tracking systems in application to small-scale photovoltaic systems. To do this, these systems are divided into two subsystems: one-axis solar tracking subsystem and two-axis solar tracking subsystem. Each is explored for its suitability as a suitable technology for small-scale application, defined by meeting a set of seven attributes. A specialized literature review process was developed specifically to present the potential of each subsystem evaluated, through an evaluation matrix. The analysis of solar tracking technologies and attributes was carried out through the formal concept analysis methodology. To facilitate the processing of information, a free access tool called concept Explorer is used. The analysis carried out shows that two-axis solar tracking systems are the most promising option as they meet a greater number of attributes than single-axis tracking systems. The applied methodology has proven to be useful for the evaluation of renewable technologies and serves as a reference, to carry out analyzes that group and link different alternatives as a model for evaluating a deterministic set of attributes.

Keywords: Solar Trackers, Formal Analysis of Concepts, Photovoltaic Systems, Renewable Energy

1. Introduction

Industrial Development and population growth accelerated the use of fossil fuels worldwide, generating an environmental crisis [1][2]. Eventually, world governments in order to combat environmental problems caused by the excessive use of fossil fuels, committed to working together to combat this problem [3]. So, from the 1970s on, renewable sources began to be considered as an alternative to traditional energies [4].

In recent decades, the field of renewable energy has been of great importance for sustainable development [5]. Currently, the use of environmentally friendly technologies for the production of electrical energy is a reality [6] [7]. Solar energy, for its part, is presented as the most abundant source worldwide, being an efficient and economical alternative, compared to other traditional forms [8] [9] [10].

Among all renewable systems, photovoltaic systems are those that present a higher degree of maturity and lead the world in the process of substitution of conventional energy resources [11] [12]. On the other hand, the solar panel is developed mainly with semiconductor materials and its maximum efficiency reaches up to 25%; with viable manufacturing and sales costs for implementation [11] [13].

Unless materials that generate high efficiency are used in photovoltaic systems, they will not be able to improve their performance [14]. This is why the viable way to improve the performance of photovoltaic panels is the application of solar trackers or trackers [16]. Solar tracking technology has been tested by different authors to increase the efficiency of solar panels and maintain their position with respect to the location of the sun during the day [11] [9] [17].

Consequently, solar trackers have been widely accepted worldwide to harness solar energy more efficiently, moving the payload of the solar panel towards the sun throughout the day [9] [18]. Solar tracking systems can be classified according to the degrees of freedom on a single axis and two axes [4] [18] [19]. Within these two classifications, the systems can vary depending on the control system used [20], the cabling used [21], tracking strategies used [22] or according to the degree of freedom of movement of the system [23]. Being innovative elements and with diverse applications for centralized and decentralized systems.

For the theoretical development of this document, the tracking systems applied to low-scale devices will be evaluated in two large groups: single-axis subsystems (1D) and two-axis subsystem (2D), because these two types of trackers have with various control systems governed by control algorithms or sensors for measuring parameters of solar radiation, to guide the position of the axes towards the position of the sun. The information analysis will be developed under the application of the formal concept analysis method [24] [25] and the use of the concept Explorer software [26], relating a series of objects and attributes, which will facilitate the processing of information. Bearing this in mind, the objectives of this work are:

i. Establish the attributes that each of the solar tracking systems to evaluate must meet
ii. Conduct a review of the academic literature of each of the solar trace systems to be evaluated under the established attributes
iii. Discuss the relationship between technology and attributes established through the use of Formal Concept Analysis

For this approach, the document is organized as follows. In section 2 the methodology is presented, defining the attributes and assigning each tracking technology an object for the development of the formal concept analysis. Section 3 covers the literary review on solar tracking systems, focusing on the level of compliance for each of the attributes. Section 4 presents the critical review using the concept Explorer tool and the formal concept analysis method, evaluating the potential of each solar tracking system to be applied to small-scale photovoltaic systems. Finally, section 5 presents the most relevant conclusions of the work.

2. Methodology

Large-scale technological developments and high investment costs often become inoperable due to the inability to maintain the initially deployed devices. For this reason, this work takes into consideration variables that consider relevant aspects for the implementation of solar tracking alternatives in low-scale photovoltaic systems. The following aspects are highlighted below:

i. Technology appreciates local know-how for assembly, operation and maintenance
ii. Technology recognizes and makes use of local resources and materials to manufacture the system

iii. The technology is sustainable, both for its long-term operation and for its maintenance

The formal analysis of concepts is based on the fulfillment to a greater or lesser extent of a series of attributes with respect to a series of objects [27]. The following are the attributes that each solar tracking system must meet:

A1-The system has technological maturity
A2-The system is easy to Operate and maintain
A3-The system presents low investment cost
A4-Widely Available Building Materials
A5-The system is efficient on a small scale
A6-The system is autonomous, does not require additional inputs (Energy use)
A7-The system is easily scalable

Considering the two solar tracking subsystems, a literature review is carried out considering the attributes. Specifically, each of the characteristics that fulfill each attribute to a lesser or greater extent. This will result in a matrix that presents the extent to which each object or solar tracking system complies with the attributes, thus allowing a qualitative analysis to be carried out.

In this work, the information evaluation is carried out using formal concept analysis (FCA), which allows data to be analyzed in terms of relationship and structure, relating them conceptually [25]. The data analysis will be carried out through the open access software "Concept Explorer" [26], evaluating all existing relationships, grouping technologies and attributes.

3. Solar Tracking Systems

Solar energy is one of the most abundant energy sources [28], this is because solar radiation has the greatest potential for clean technologies since it is a clean, free, abundant and inexhaustible source of energy [29]. Solar energy in recent years has been used by fixed or static solar systems, but with the development of different types of technology, the efficiency of static systems has been increasing through the use of solar trackers[23].

Solar tracking systems can be classified according to the degrees of freedom on a single axis and two axes. Within these two classifications, the systems can vary depending on the control system used [20], the cabling used [21], tracking strategies used [22] or according to the degree of freedom of movement of the system [23]. For this project, systems with one degree of freedom and two degrees of freedom will be evaluated, including the subclassifications that each one may have with the characteristics mentioned above [30].

It is important to note that the degree of freedom represents the number of directions in which the system can move. Tracking axes play an imperative role in each specific application and determine the proper location and directions of the solar panel [23].

Table 1 presents the classification of trackers to evaluate:

Table 1. Solar trackers classification

Technology	Subsystem
Based on degrees of movement	Single axis Two-axis

Source: Author

Next, the subsystems of the solar tracking technologies presented in the Table 1.

3.1. Single Axis Solar Tracking Subsystems (1D)

This system involves a single axis of movement, around which the solar panel rotates perpendicular to the sun's radiation and is considered a mature technology at the laboratory and commercial level [31] [23]. There are several configurations with this type of tracker including vertical alignment (V1D)[32], horizontal alignment (H1D) [33], horizontal alignment with inclined modules (H1D) [34], inclined alignment (T1D) and polar alignment (P1D) [35].

The main components of the system generally require a mechanism and a transmission system, as well as a motor and a control system, governed by microprocessors, which include the development of solar algorithms or treatment of data from sensors that measure solar radiation [36] [37]. It should be noted that components for large, medium and low-scale systems can be obtained locally at the cost of increasing the solar tracking error compared to commercial systems; depending less on technological maturity and more on the size of the photovoltaic system [38] [39].

The 1D solar tracking system is the simplest and most economical Tracker [40] [41], however its efficiency is low due to the fact that under operating conditions, the photovoltaic module can be directed only horizontally or vertically [42] [43]. As efficiency is generally less than two-axis tracking systems, the technology is less mature and commercial developments are more limited [44] [45].

The energy consumption requirements of these systems correspond clearly to the power supply of the microprocessor, as well as the consumption of the motor and the radiation sensor (s) used [44] [46] [47]. In turn, the maintenance and operation of this type of system mainly involves the change and / or relationship of parts, as well as technical knowledge in programming [23] [48] [49].

Finally, the modular nature of this type of system allows its development at a centralized and decentralized level [23] [44] [50]. At the laboratory level, there are a large number of small-scale prototypes developed, tested and finally scaled to commercial sizes, to be implemented in solar fields at a centralized scale [34] [51].

3.2. Two Axis Solar Tracking Subsystem (2D)

This system is based on two axes of rotation that are generally perpendicular to each other [52] [53]. Generally, the system moves in azimuth and elevation, and currently has countless commercial and industrial applications [54], with solar and satellite tracking applications [55] [56] [57].

The main components of the two-axis tracking system (2D) are similar to the components of the 1D system, they differ in the number of elements, that is, in the case of the 2D system, the tracking mechanism must integrate two or more tracking systems. transmission, as well as two independent motors for each axis and a control system, governed by a set of microcontrollers, which will involve further development of software (control algorithms) [57] [58] or application of solar radiation sensors [59] [60]. As in 1D systems, the components of 2D tracking devices can be applied on a large, medium or small scale, they can be purchased or manufactured locally; its impact on the efficiency of the system will be seen, depending on the degree of appropriate technology used for its developmen [61] [62].

On the other hand, the 2D tracking system is more complex in terms of construction and programming than the 1D tracking system [63]. Consequently, its investment cost is higher, however, its efficiency is high because, under photovoltaic system operating conditions, it follows the solar path horizontally and vertically at the same time [64]. Additionally, by presenting greater efficiency than the 1D system, its level of technological development is higher, generating a large number of technological applications worldwide [65] [66].

The consumption requirements of this system correspond in the same way as the 1D system to the supply of electronic components (microcontrollers, motors, etc.) [67]. It is important to note that the power consumption in the case of 2D trackers is higher, due to the increase in system components [68]. in turn, maintenance and operation basically requires the change of spare parts, as well as technical programming knowledge, regardless of whether it has a higher degree of complexity than 1D systems, due to the number of elements [23] [44] [69].

Finally, the 2D solar tracking system presents development at a centralized and decentralized level worldwide [44] [70]. Its development has low-scale systems developed in laboratories and implemented in buildings [23] [71], as well as medium and large-scale systems, installed in solar farms and plants for the massive production of electricity through the use of solar panels [72] [73].

IOP Conf. Series: Materials Science and Engineering 1253 (2022) 012002 doi:10.1088/1757-899X/1253/1/012002

Table 2 presents the attribute matrix for the two solar tracking technologies. The attributes that meet for each technologist are identified with an X within the matrix.

Table 2. Selection matrix of attributes with respect to objects

At / Ob	A1	A2	A3	A4	A5	A6	A7
1D			X	X		X	X
2D	X		X	X	X	X	X

Source: Author

4. Results

Figure 1 represents the graphic diagram of the formal analysis of concepts carried out in the concept Explorer tool for the analysis of the 1D and 2D solar trackers systems, based on the information consigned in the attribute matrix of Table 2. This diagram relates the solar trackers of one axis (1D) and two-axis solar trackers (2D) with those 7 attributes proposed to identify the application of solar tracking technologies to low-scale photovoltaic systems.

The different nodes shown in Figure 1 are identified as follows:

i. White-blue: contains only attributes
ii. Blue-black - Containers of attribute-related solar tracking technologies

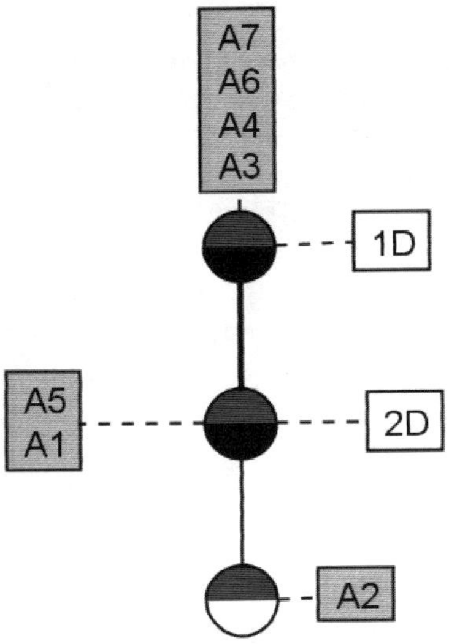

Figure 1. Information analysis diagram

Considering the above, the analysis of the diagram is simple. The two analyzed solar tracking technologies are related to all the attributes that connect it from top to bottom, to the node of origin. Specifically, the attribute “A2-the system is easy to operate and maintain” is not fulfilled by any of the “Types of solar trackers” objects. For its part, the two-axis solar tracking system (2D) complies with two more attributes than the single-axis tracking system (1D), that is, the 2D system has technological maturity "A1" and is efficient at low scale (A5), while the 1D system does not. On the other hand, both the 1D and 2D systems meet the following attributes at the same time: the system has a low investment cost "A3", the construction materials are widely available "A4", the system is autonomous "A6" and the system is easily scalable "A7".

From this analysis, it is evident that the two solar tracking technologies have sought to provide solutions at a centralized and decentralized level, to improve the efficiency of photovoltaic solar systems that are installed in a static way. As an alternative to the theoretical analysis approach, we can affirm that the two-axis (2D) solar tracking technology is the most suitable for applications in small-scale photovoltaic systems. However, single axis (1D) tracking technology can overcome the attributes that it does not meet, through the development of new systems in the short and medium term.

5. Conclusions

Solar tracking systems will play an important role in increasing the efficiency of solar systems around the world, but it is still necessary to carry out massive implementations of centralized and decentralized photovoltaic systems, especially in developing countries. This type of technological development will be accompanied by solar tracking systems that will generate a positive impact on the generation of electrical energy. At the same time, the use of local resources for its development will offer a sustainable solution for its implementation, overcoming social and economic limitations present in this specific type of communities.

Given the large amounts of radiation present in most territories worldwide, the use of solar trackers as an alternative for their implementation in low-scale photovoltaic systems is analyzed, applying the methodology of formal analysis of concepts, adapted for this purpose. The tracking systems were divided in two; one-axis (1D) and two-axis (2D). From the theoretical analysis of these two systems the following conclusions were obtained:

i. The solar tracking systems were grouped into two groups, single-axis 1D and two-axis 2D systems, since these systems include control characteristics, software development and use of sensorics for solar tracking, which in some review documents are independently analyzed.

ii. Of the two large groups of solar tracking technologies analyzed, the two-axis (2D) systems present better characteristics based on the analyzed attributes, to be adopted by low-scale photovoltaic systems, with the use of local resources, maintaining the system efficiency at a profitable cost.

iii. Single axis (1D) systems are simpler and cheaper systems, but they lack the high efficiency of 2D systems, which somewhat limits their application in both centralized and decentralized systems.

Finally, the academic and scientific information analyzed allows credibility to the quantitative analysis developed, although this type of analysis should not replace qualitative analysis, it appears as an alternative to conventional studies.

References

[1] R. O. A. Moscoso, «La Industria y sus efectos en el cambio climático Global», *RECIAMUC*, vol. 2, n.º 2, Art. n.º 2, 2018, doi: 10.26820/reciamuc/2.2.2018.595-611.

[2] B. E. Tarazona-Romero, A. Campos-Celador, Y. A. Muñoz-Maldonado, J. G. Ascanio-Villabona, M. A. Duran-Sarmiento, y A. D. Rincón-Quintero, «Development of a Fresnel Artisanal System for the Production of Hot Water or Steam», en *Recent Advances in Electrical Engineering, Electronics and Energy*, Cham, 2021, pp. 196-209. doi: 10.1007/978-3-030-72212-8_15.

[3] D. Gielen, F. Boshell, D. Saygin, M. D. Bazilian, N. Wagner, y R. Gorini, «The role of renewable energy in the global energy transformation», *Energy Strategy Reviews*, vol. 24, pp. 38-50, abr. 2019, doi: 10.1016/j.esr.2019.01.006.

[4] N. Machado Toranzo, A. Lussón Cervantes, L. L. Oro Carralero, J. Bonzon Henríquez, y O. Escalona Costa, «Seguidor Solar, optimizando el aprovechamiento de la energía solar», *Ingeniería Energética*, vol. 36, n.º 2, pp. 190-199, ago. 2015.

[5] M. Shahbaz, C. Raghutla, K. R. Chittedi, Z. Jiao, y X. V. Vo, «The effect of renewable energy consumption on economic growth: Evidence from the renewable energy country attractive index», *Energy*, vol. 207, p. 118162, sep. 2020, doi: 10.1016/j.energy.2020.118162.

[6] R. Vakulchuk, I. Overland, y D. Scholten, «Renewable energy and geopolitics: A review», *Renewable and Sustainable Energy Reviews*, vol. 122, p. 109547, abr. 2020, doi: 10.1016/j.rser.2019.109547.
[7] B. E. T. Tarazona-Romero, Á. C. Celador, Y. A. M. Maldonado, C. S. Rodríguez, y J. G. A. Villabona, «Prototype of lineal solar collector Fresnel», *Visión electrónica*, vol. 14, n.º 1, p. 4, 2020.
[8] B. E. Tarazona-Romero, A.-J. R. Nieves, C. L. Sandoval-Rodriguez, J. G. Ascanio-Villabona, M.-A. Duran-Sarmiento, y A.-D. Rincon-Quintero, «Sizing of a sustainable condensation system for the production of sweet water through the use of photovoltaic panels in La Mesa de los Santos, Santander, Colombia», *IOP Conf. Ser.: Mater. Sci. Eng.*, vol. 1154, n.º 1, p. 012040, jun. 2021, doi: 10.1088/1757-899X/1154/1/012040.
[9] S. Racharla y K. Rajan, «Solar tracking system – a review», *International Journal of Sustainable Engineering*, vol. 10, n.º 2, pp. 72-81, mar. 2017, doi: 10.1080/19397038.2016.1267816.
[10] H. Wang, Z. Lei, X. Zhang, B. Zhou, y J. Peng, «A review of deep learning for renewable energy forecasting», *Energy Conversion and Management*, vol. 198, p. 111799, oct. 2019, doi: 10.1016/j.enconman.2019.111799.
[11] Md. T. A. Khan, S. M. S. Tanzil, R. Rahman, y S. M. S. Alam, «Design and construction of an automatic solar tracking system», en *International Conference on Electrical Computer Engineering (ICECE 2010)*, dic. 2010, pp. 326-329. doi: 10.1109/ICELCE.2010.5700694.
[12] A. Taheri *et al.*, «Improving the performance of a nanofluid-based photovoltaic thermal module utilizing dual-axis solar tracker system: Experimental examination and thermodynamic analysis», *Applied Thermal Engineering*, vol. 196, p. 117178, sep. 2021, doi: 10.1016/j.applthermaleng.2021.117178.
[13] S. Makhdoomi y A. Askarzadeh, «Impact of solar tracker and energy storage system on sizing of hybrid energy systems: A comparison between diesel/PV/PHS and diesel/PV/FC», *Energy*, vol. 231, p. 120920, sep. 2021, doi: 10.1016/j.energy.2021.120920.
[14] S. Z. Golroodbari y W. van Sark, «Simulation of performance differences between offshore and land-based photovoltaic systems», *Progress in Photovoltaics: Research and Applications*, vol. 28, n.º 9, pp. 873-886, 2020, doi: 10.1002/pip.3276.
[15] A. K. Bhoi, P. K. Mallick, V. E. Balas, y B. S. P. Mishra, *Advances in Systems, Control and Automations: Select Proceedings of ETAEERE 2020*. Springer Nature, 2021.
[16] W. Manosroi, P. Prompattra, y P. Kerngburee, «Performance improvement of two-axis solar tracking system by using flat-mirror reflectors», *Energy Reports*, vol. 6, pp. 9-14, dic. 2020, doi: 10.1016/j.egyr.2020.10.029.
[17] L. F. Saavedra Quintero y D. F. Sánchez Martínez, «Automatización de un sistema de seguimiento solar en un eje para un prototipo de colector lineal tipo Fresnel de las Unidades Tecnológicas de Santander», abr. 2021, Accedido: jul. 10, 2021. [En línea]. Disponible en: http://repositorio.uts.edu.co:8080/xmlui/handle/123456789/5828
[18] M. E. H. Chowdhury, A. Khandakar, B. Hossain, y R. Abouhasera, «A Low-Cost Closed-Loop Solar Tracking System Based on the Sun Position Algorithm», *Journal of Sensors*, vol. 2019, p. e3681031, feb. 2019, doi: 10.1155/2019/3681031.
[19] M. Abdollahpour, M. R. Golzarian, A. Rohani, y H. Abootorabi Zarchi, «Development of a machine vision dual-axis solar tracking system», *Solar Energy*, vol. 169, pp. 136-143, jul. 2018, doi: 10.1016/j.solener.2018.03.059.
[20] H. J. Loschi, Y. Iano, J. León, A. Moretti, F. D. Conte, y H. Braga, «A review on photovoltaic systems: mechanisms and methods for irradiation tracking and prediction», 2015, doi: 10.4236/sgre.2015.67017.
[21] A. Elmaged y H. Kamal, «Passive Solar Tracking System», Thesis, UOFK, 2015. Accedido: jul. 16, 2021. [En línea]. Disponible en: https://khartoumspace.uofk.edu/handle/123456789/13328

[22] I. Chaidoulis y N. N. Karanikolas, «A review of microprocessor based solar photovoltaic tracking systems», en *24th Pan-Hellenic Conference on Informatics*, New York, NY, USA, nov. 2020, pp. 82-86. doi: 10.1145/3437120.3437280.
[23] A. Awasthi *et al.*, «Review on sun tracking technology in solar PV system», *Energy Reports*, vol. 6, pp. 392-405, nov. 2020, doi: 10.1016/j.egyr.2020.02.004.
[24] B. Ganter y R. Wille, *Análisis de concepto formal, Texto original: Fundamentos matemáticos*. Berlín Heidelberg: Springer-Verlag, 1999. Accedido: jul. 09, 2021. [En línea]. Disponible en: https://www.springer.com/gp/book/9783540627715
[25] F. Sancho Caparrini, «Análisis Formal de Conceptos», *Análisis Formal de Conceptos*, 2017. http://www.cs.us.es/~fsancho/?e=78 (accedido jul. 08, 2021).
[26] S. Yevtushenko, J. Tane, T. B. Kaiser, S. Objedkov, J. Hereth Correia, y H. Reppe, *Concept Explorer*. 2021. [En línea]. Disponible en: http://conexp.sourceforge.net/download.html
[27] V. Varga, K. Tünde, y K. T. Jánosi-Rancz, *A Software Tool to Transform Relational Databases in order to Mine Functional Dependencies in it Using Formal Concept Analysis*. 2008.
[28] G. Li, S. Shittu, T. M. O. Diallo, M. Yu, X. Zhao, y J. Ji, «A review of solar photovoltaic-thermoelectric hybrid system for electricity generation», *Energy*, vol. 158, pp. 41-58, sep. 2018, doi: 10.1016/j.energy.2018.06.021.
[29] A. F. de Paulo y G. S. Porto, «Evolution of collaborative networks of solar energy applied technologies», *Journal of Cleaner Production*, vol. 204, pp. 310-320, dic. 2018, doi: 10.1016/j.jclepro.2018.08.344.
[30] A. Z. Hafez, A. M. Yousef, y N. M. Harag, «Solar tracking systems: Technologies and trackers drive types – A review», *Renewable and Sustainable Energy Reviews*, vol. 91, pp. 754-782, ago. 2018, doi: 10.1016/j.rser.2018.03.094.
[31] S. Avarand y M. Pirmoradian, «Solar Tracking System with Momentary Tracking Based on Operational Amplifiers in Order to be Used in Photovoltaic Panels for Following the Sun», *Bulletin de la Société Royale des Sciences de Liège*, vol. 85, p. 9, 2016.
[32] L. Li, H. Li, Q. Xu, y W. Huang, «Performance analysis of Azimuth Tracking Fixed Mirror Solar Concentrator», *Renewable Energy*, vol. 75, pp. 722-732, mar. 2015, doi: 10.1016/j.renene.2014.10.062.
[33] B. A. D. J. C. K. Basnayake, W. A. D. M. Jayathilaka, Y. W. R. Amarasinghe, R. A. Attalage, y A. G. B. P. Jayasekara, «Smart solar tracking and on-site photovoltic efficiency measurement system», en *2016 Moratuwa Engineering Research Conference (MERCon)*, abr. 2016, pp. 54-59. doi: 10.1109/MERCon.2016.7480115.
[34] A. S. C. Roong y S.-H. Chong, «Laboratory-Scale Single Axis Solar Tracking System: Design and Implementation», *International Journal of Power Electronics and Drive Systems (IJPEDS)*, vol. 7, n.° 1, Art. n.° 1, mar. 2016, doi: 10.11591/ijpeds.v7.i1.pp254-264.
[35] G. C. Lazaroiu, M. Longo, M. Roscia, y M. Pagano, «Comparative analysis of fixed and sun tracking low power PV systems considering energy consumption», *Energy Conversion and Management*, vol. 92, pp. 143-148, mar. 2015, doi: 10.1016/j.enconman.2014.12.046.
[36] M. S. AlHammad, «A Hybrid Photodiode for Solar Tracking Systems», *Silicon*, vol. 13, n.° 1, pp. 243-250, ene. 2021, doi: 10.1007/s12633-020-00423-3.
[37] M. Das y E. K. Akpinar, «Investigation of the effects of solar tracking system on performance of the solar air dryer», *Renewable Energy*, vol. 167, pp. 907-916, abr. 2021, doi: 10.1016/j.renene.2020.12.010.
[38] A. A. Elsayed, E. E. Khalil, M. A. Kassem, y O. A. Huzzayin, «A novel mechanical solar tracking mechanism with single axis of tracking for developing countries», *Renewable Energy*, vol. 170, pp. 1129-1142, jun. 2021, doi: 10.1016/j.renene.2021.02.058.
[39] A. Z. Hafez, A. M. Yousef, y N. M. Harag, «Solar tracking systems: Technologies and trackers drive types – A review», *Renewable and Sustainable Energy Reviews*, vol. 91, pp. 754-782, ago. 2018, doi: 10.1016/j.rser.2018.03.094.

[40]R. Bakhshi y J. Sadeh, «A comprehensive economic analysis method for selecting the PV array structure in grid–connected photovoltaic systems», *Renewable Energy*, vol. 94, pp. 524-536, ago. 2016, doi: 10.1016/j.renene.2016.03.091.
[41]A. Bahrami, C. O. Okoye, y U. Atikol, «Technical and economic assessment of fixed, single and dual-axis tracking PV panels in low latitude countries», *Renewable Energy*, vol. 113, pp. 563-579, dic. 2017, doi: 10.1016/j.renene.2017.05.095.
[42]N. AL-Rousan, N. A. M. Isa, y M. K. M. Desa, «Advances in solar photovoltaic tracking systems: A review», *Renewable and Sustainable Energy Reviews*, vol. 82, pp. 2548-2569, feb. 2018, doi: 10.1016/j.rser.2017.09.077.
[43]N. AL-Rousan, N. A. Mat Isa, y M. K. Mat Desa, «Efficient single and dual axis solar tracking system controllers based on adaptive neural fuzzy inference system», *Journal of King Saud University - Engineering Sciences*, vol. 32, n.º 7, pp. 459-469, nov. 2020, doi: 10.1016/j.jksues.2020.04.004.
[44]A. R. Amelia, Y. M. Irwan, I. Safwati, W. Z. Leow, M. H. Mat, y M. S. A. Rahim, «Technologies of solar tracking systems: A review», 2020, vol. 767, n.º 1. doi: 10.1088/1757-899X/767/1/012052.
[45]M. T. Patel, M. S. Ahmed, H. Imran, N. Z. Butt, M. R. Khan, y M. A. Alam, «Global analysis of next-generation utility-scale PV: Tracking bifacial solar farms», *Applied Energy*, vol. 290, p. 116478, may 2021, doi: 10.1016/j.apenergy.2021.116478.
[46]H. Wang, T. Luo, Y. Fan, Z. Lu, H. Song, y J. B. Christen, «A self-powered single-axis maximum power direction tracking system with an on-chip sensor», *Solar Energy*, vol. 112, pp. 100-107, feb. 2015, doi: 10.1016/j.solener.2014.11.019.
[47]R. Shaik, N. Beemkumar, H. Adharsha, K. Venkadeshwaran, y A. D. Dhass, «Efficiency enhancement in a PV operated solar pump by effective design of VFD and tracking system», *Materials Today: Proceedings*, vol. 33, pp. 454-462, ene. 2020, doi: 10.1016/j.matpr.2020.05.035.
[48]P. Munanga, S. Chinguwa, W. R. Nyemba, y C. Mbohwa, «Design for manufacture and assembly of an intelligent single axis solar tracking system», *Procedia CIRP*, vol. 91, pp. 571-576, ene. 2020, doi: 10.1016/j.procir.2020.03.109.
[49]J. A. Forero Monsalve y B. A. Jaimes Grimaldos, «Desarrollo de un sistema automatizado de seguimiento del sol para dar movimiento a un colector solar cilindro parabólico a través de la implementación de un chip Esp32 y un Servomotor monitoreados por un miniordenador Raspberry», ene. 2021, Accedido: jul. 10, 2021. [En línea]. Disponible en: http://repositorio.uts.edu.co:8080/xmlui/handle/123456789/5182
[50]R. Singh, S. Kumar, A. Gehlot, y R. Pachauri, «An imperative role of sun trackers in photovoltaic technology: A review», *Renewable and Sustainable Energy Reviews*, vol. 82, pp. 3263-3278, feb. 2018, doi: 10.1016/j.rser.2017.10.018.
[51]M. A. Vaziri Rad, A. Toopshekan, P. Rahdan, A. Kasaeian, y O. Mahian, «A comprehensive study of techno-economic and environmental features of different solar tracking systems for residential photovoltaic installations», *Renewable and Sustainable Energy Reviews*, vol. 129, p. 109923, sep. 2020, doi: 10.1016/j.rser.2020.109923.
[52]C. Chen, P. Li, y X. Ping, «Robust Variance Control Algorithm for the Dish Solar Generation Tracking System», en *Proceedings of the Second International Conference on Mechatronics and Automatic Control*, Cham, 2015, pp. 237-245. doi: 10.1007/978-3-319-13707-0_27.
[53]M. Ghassoul, «A dual solar tracking system based on a light to frequency converter using a microcontroller», *Fuel Communications*, vol. 6, p. 100007, mar. 2021, doi: 10.1016/j.jfueco.2020.100007.
[54]H. Fathabadi, «Comparative study between two novel sensorless and sensor based dual-axis solar trackers», *Solar Energy*, vol. 138, pp. 67-76, nov. 2016, doi: 10.1016/j.solener.2016.09.009.
[55]F. Sallaberry, R. Pujol-Nadal, M. Larcher, y M. H. Rittmann-Frank, «Direct tracking error characterization on a single-axis solar tracker», *Energy Conversion and Management*, vol. 105, pp. 1281-1290, nov. 2015, doi: 10.1016/j.enconman.2015.08.081.

[56] J. Sun, R. Wang, H. Hong, y Q. Liu, «An optimized tracking strategy for small-scale double-axis parabolic trough collector», *Applied Thermal Engineering*, vol. 112, pp. 1408-1420, feb. 2017, doi: 10.1016/j.applthermaleng.2016.10.187.

[57] M. H. M. Sidek, N. Azis, W. Z. W. Hasan, M. Z. A. Ab Kadir, S. Shafie, y M. A. M. Radzi, «Automated positioning dual-axis solar tracking system with precision elevation and azimuth angle control», *Energy*, vol. 124, pp. 160-170, abr. 2017, doi: 10.1016/j.energy.2017.02.001.

[58] M. Ali Jallal, S. Chabaa, y A. Zeroual, «A novel deep neural network based on randomly occurring distributed delayed PSO algorithm for monitoring the energy produced by four dual-axis solar trackers», *Renewable Energy*, vol. 149, pp. 1182-1196, abr. 2020, doi: 10.1016/j.renene.2019.10.117.

[59] M.-H. Tan *et al.*, «Optimization Study of Parasitic Energy Losses in Photovoltaic System with Dual-Axis Solar Tracker Located at Different Latitudes», *Energy Procedia*, vol. 158, pp. 302-308, feb. 2019, doi: 10.1016/j.egypro.2019.01.093.

[60] W. Nsengiyumva, S. G. Chen, L. Hu, y X. Chen, «Recent advancements and challenges in Solar Tracking Systems (STS): A review», *Renewable and Sustainable Energy Reviews*, vol. 81, pp. 250-279, ene. 2018, doi: 10.1016/j.rser.2017.06.085.

[61] H. Fathabadi, «Novel high accurate sensorless dual-axis solar tracking system controlled by maximum power point tracking unit of photovoltaic systems», *Applied Energy*, vol. 173, pp. 448-459, jul. 2016, doi: 10.1016/j.apenergy.2016.03.109.

[62] E. K. Mpodi, Z. Tjiparuro, y O. Matsebe, «Review of dual axis solar tracking and development of its functional model», *Procedia Manufacturing*, vol. 35, pp. 580-588, ene. 2019, doi: 10.1016/j.promfg.2019.05.082.

[63] A. Saymbetov *et al.*, «Dual-axis schedule tracker with an adaptive algorithm for a strong scattering of sunbeam», *Solar Energy*, vol. 224, pp. 285-297, ago. 2021, doi: 10.1016/j.solener.2021.06.024.

[64] Y. M. Safan, S. Shaaban, y M. I. Abu El-Sebah, «Performance evaluation of a multi-degree of freedom hybrid controlled dual axis solar tracking system», *Solar Energy*, vol. 170, pp. 576-585, ago. 2018, doi: 10.1016/j.solener.2018.06.011.

[65] T. Salameh, C. Ghenai, A. Merabet, y M. Alkasrawi, «Techno-economical optimization of an integrated stand-alone hybrid solar PV tracking and diesel generator power system in Khorfakkan, United Arab Emirates», *Energy*, vol. 190, p. 116475, ene. 2020, doi: 10.1016/j.energy.2019.116475.

[66] T. Rajesh, K. S. Tamilselvan, A. Vijayalakshmi, Ch. Narendra Kumar, y K. Anitha Reddy, «Design and implementation of an automatic solar tracking system for a monocrystalline silicon material panel using MPPT algorithm», *Materials Today: Proceedings*, vol. 45, pp. 1783-1789, ene. 2021, doi: 10.1016/j.matpr.2020.08.635.

[67] Y. Zhu, J. Liu, y X. Yang, «Design and performance analysis of a solar tracking system with a novel single-axis tracking structure to maximize energy collection», *Applied Energy*, vol. 264, p. 114647, abr. 2020, doi: 10.1016/j.apenergy.2020.114647.

[68] A. Bahrami y C. O. Okoye, «The performance and ranking pattern of PV systems incorporated with solar trackers in the northern hemisphere», *Renewable and Sustainable Energy Reviews*, vol. 97, pp. 138-151, dic. 2018, doi: 10.1016/j.rser.2018.08.035.

[69] V. Sumathi, R. Jayapragash, A. Bakshi, y P. Kumar Akella, «Solar tracking methods to maximize PV system output – A review of the methods adopted in recent decade», *Renewable and Sustainable Energy Reviews*, vol. 74, pp. 130-138, jul. 2017, doi: 10.1016/j.rser.2017.02.013.

[70] S. Yilmaz, H. Riza Ozcalik, O. Dogmus, F. Dincer, O. Akgol, y M. Karaaslan, «Design of two axes sun tracking controller with analytically solar radiation calculations», *Renewable and Sustainable Energy Reviews*, vol. 43, pp. 997-1005, mar. 2015, doi: 10.1016/j.rser.2014.11.090.

[71] N. AL-Rousan, N. A. M. Isa, y M. K. M. Desa, «Advances in solar photovoltaic tracking systems: A review», *Renewable and Sustainable Energy Reviews*, vol. 82, pp. 2548-2569, feb. 2018, doi: 10.1016/j.rser.2017.09.077.

[72]B.-H. Lim *et al.*, «Industrial design and implementation of a large-scale dual-axis sun tracker with a vertical-axis-rotating-platform and multiple-row-elevation structures», *Solar Energy*, vol. 199, pp. 596-616, mar. 2020, doi: 10.1016/j.solener.2020.02.006.
[73]X. Du, Y. Li, P. Wang, Z. Ma, D. Li, y C. Wu, «Design and optimization of solar tracker with U-PRU-PUS parallel mechanism», *Mechanism and Machine Theory*, vol. 155, p. 104107, ene. 2021, doi: 10.1016/j.mechmachtheory.2020.104107.

Incident factors in the quality of cocoa almonds obtained in small agricultural production units

E. L. Grimaldo-Pretel[1], M. A. Guerrero-Navarro[1], C.R. Vidal-Tovar[1], Y. Gordon-Hernández[2] and P. J. Fragoso-Castilla[3]

1 Grupo de Investigación en Sistemas Socio productivos – Universidad Popular del Cesar. Valledupar, Cesar, Colombia.
2. Grupo FACEUPC, Universidad Popular del Cesar. Universidad Popular del Cesar. Valledupar, Cesar, Colombia.
3 Grupo de Investigación en Parasitologia - Agroecologia Milenio Universidad Popular del Cesar. Universidad Popular del Cesar. Valledupar, Cesar, Colombia.
carlosvidal@unicesar.edu.co

Abstract. Cocoa is an American tree of Amazonian origin, also known as cacaotero, since by cocoa it usually refers to the fruit that gives said tree. The proposed research is developed on the emerging paradigm, for which a dialectical synthesis is assumed from the use of qualitative and quantitative methods in a complementary way. Not limited to the forms and sources of information the collection of information, and the analysis of results, for which we take as a reference the contribution from the experience of Cocoa Producers and Experts in the production and marketing of cocoa, as a fundamental part of the construction of knowledge, and grant the same level of importance to the information obtained from multiple sources of information obtained in the year 2020. For the collection of information, the instruments are applied in three phases: the first consists of field visits to interview 5 cocoa producers and 5 experts in cocoa marketing. The second, search and systematization of documents and the third phase is a comparative analysis of the sources of information mentioned above. For the establishment of cocoa cultivation and to obtain adequate productions, aspects of great relevance must be taken into account, such as critical climatic factors such as temperature and rain. The layout and density of the agroforestry system. This activity aims to make a defined and organized distribution within the lot of each of the plant species that are going to be planted, seeking a better use of the soil. It was verified the effort made by small producers to integrate in the management of their cocoa crops, the processes, harvesting and post-harvest activities that are carried out to obtain dried cocoa almonds for marketing.

Keywords: Cocoa, Fermentation, Almonds, Chocolate Quality.

1. Introduction

Cocoa " is an American tree of Amazonian origin, also known as cacaotero, since by cocoa it usually refers to the fruit that gives said tree, or even to the product of the drying and fermentation of the seeds of said fruit" [1]. According to the National Federation of Cocoa Farmers "During the last ten years Colombia has shown a significant growth in cocoa production, going from 36,118 tons in 2009 to 59,740 in 2019; in addition to a significant increase in exports and a notable decrease in imports of grain for domestic consumption" [2]. In addition, there is evidence of a significant support in new planting and renewal of the crop in the national territory, because the work promoted by the National Federation of Cocoa, who has had discussions with entities such as the Ministry of Agriculture and Rural Development, to solicit the support of the renewal initial 4,000 hectares of cocoa aged, which will allow the sub to be more productive!"[3].

In this research will be conducted to determine the factors that affect the quality of cocoa almonds produced in small producers located in the towns of San José de Oriente and Media Luna in the department of Cesar, Colombia, which are in a process of growth and improvement of quality in its different stages to obtain cocoa almonds according to the requirements for marketing. This research seeks to provide a series of knowledge and technical resources that generate optimal management of cocoa crops for obtaining dried cocoa almonds in small productive units [4], [5], [6].

2. Material and methods

The proposed research is developed on the emerging paradigm, for which a dialectical synthesis is assumed from the use of qualitative and quantitative methods in a complementary way. Not limited to the forms and sources of information the collection of information, and the analysis of results, for which we take as a reference the contribution from the experience of Cocoa Producers and Experts in the production and marketing of cocoa, as a fundamental part of the construction of knowledge, and grant the same level of importance to the information obtained from multiple sources of information obtained in the year 2020 in the search engines like: Scielo, Dialnet and Google Scholar, among others. Documents from authors from different countries were reviewed through theses, books, articles, and studies, which allowed a critical reading and objective analysis on cocoa production, post-harvest processing and its influence on the quality of dried almonds for marketing.

For the collection of information, the instruments are applied in three phases: the first consists of field visits to interview 5 cocoa producers and 5 experts in cocoa marketing. The second, search and systematization of documents and the third phase is a comparative analysis of the sources of information mentioned above. The data analysis was carried out through the structuring of thematic or conceptual categories from the order of importance of factors related to the quality of cocoa provided by producers and experts. With the above, the results of this work are ordered, interpreted and presented through what has been proposed by multiple authors [7], [8], [9], and [10].

3 Results and Discussion

The results below are presented from the contributions given by the small cocoa producers, the experts consulted and the comparison, verification with various authors on the socialized topics.

3.1 Climate and soil needs for Cocoa production

For the establishment of cocoa cultivation and to obtain adequate productions, aspects of great relevance must be taken into account, such as critical climatic factors such as temperature and rain. These are joined by wind and light or solar radiation. Cocoa is a plant that grows under shade. Relative humidity is also important as it can contribute to the spread of some fruit diseases. These climatic requirements have led to cocoa cultivation being concentrated in the tropical lowlands [11], [12], [13], and [14].

Temperature. The cultivation of cocoa does not withstand low temperatures, being its average annual temperature limit of 21 ºC since it is difficult to grow cocoa satisfactorily with a lower temperature. Very high extreme temperatures can cause physiological alterations in the tree so it is a crop that must be under shade so that the sun's rays do not affect directly and the temperature increases [11], [12], [13], and [14]. The temperature determines the formation of flowers. When it is less than 21 ºC the flowering is less than 25 ºC, where the flowering is normal and abundant. This causes that in certain areas the production of cobs is seasonal and for some weeks there is no harvest, when temperatures are below 22 ºC"

Water. The water in the cultivation of cocoa being "a plant sensitive to scarcity, but also to stagnation so that soils provided with a good drainage will be needed. A waterlogging or stagnation can lead to suffocation of the roots and their death in a very short time. Water requirements range from 1500 to

2500 mm in the warmer lowlands and from 1200 to 1500 mm in the cooler areas or high valleys" [11], [12], [13], and [14].

Wind. The wind according to several sources if these are continuous "can cause a drying, death and fall of the leaves. In coastal areas, therefore, windbreaks are needed to ensure that cocoa is not damaged. Windbreaks are usually formed by different tree species (fruit or wood) that are arranged around cocoa trees" [11], [12], [13], and [14].

Shading. For this stage "the objective of shading at the beginning of the plantation is to reduce the amount of radiation that reaches the crop to reduce the activity of the plant and protect the crop from the winds that can damage it. When the crop is established, the percentage of shading can be reduced to 25 or 30 %. The luminosity must be more or less 50% during the first 4 years of life of the plants, so that they reach a good development and limit the growth of bad herbs. For the shade of the crop are used the so-called shade species, which are usually other fruit trees interspersed in the crop with regular planting frames. The most used species are the musaceae (plantain, topochos and cambures) for temporary shadows and legumes such as poró or bucare (Erythrina sp.) and the guabas (Ingas) for permanent shadows. New cocoa plantations are beginning to use other shading species that provide greater economic benefit such as timber (laurel, cedar, cenizaro and terminalia) and/or fruit (citrus, avocado, sapote, bread tree, date palm, etc.)" [11], [12], [13], and [14].

Ground demands. Finally, "Cocoa requires very rich soils in organic matter, deep, loamy loam, with good drainage and regular topography. The limiting factor of the soil in the development of cocoa is the thin humic layer. This layer degrades very quickly when the soil surface is exposed to sun, wind and direct rain. Therefore it is common to use auxiliary leguminous plants that provide the necessary shade and are a constant source of nitrogen substances for cultivation. The plantations are located in soils ranging from heavily eroded heavy clays to newly formed volcanic sands and silts, with pH ranging between 4.0 and 7.0. It can be said that cocoa is a plant that thrives in a wide diversity of soil types" [11], [12], [13], and [14].

3.2 Characteristics and support operations for cocoa production.

The layout and density of the agroforestry system. This activity aims to make a defined and organized distribution within the lot of each of the plant species that are going to be planted, seeking a better use of the soil. Steps for the establishment of the agroforestry system: First, the layout is made for the cocoa trees, using the triangle planting system, locating sites for the hollowing every 3 meters between sites, obtaining a density between 900 to 1100 cocoa trees per hectare, which depend on the planting distance and the agro-forestry system applied. This work is carried out with the East – West direction in mind. Subsequently, in the same sense of the path for cocoa is made the path for the transitory shade (banana, banana, passion fruit, fig, papaya, yucca, guandul, others), placing a mark at the midpoint corresponding to the distance between two cocoa trees, the sites of the transitory shade are also marked every 3 meters each other, and a density of 1,282 plants of transitory shade is obtained [15], [16], [17], [18], and [19].

Finally, the path in barriers for timber trees is made in the South-North direction, tracing points at a distance of 5 meters between trees, and 15 meters between the paths of timber trees. The number of trees per hectare depends on the tree species being planted, the distance between trees, and the distance between tree rows. In the case of this example, with these planting distances, 133 trees per hectare are obtained. In order to provide additional coverage, between the cocoa and banana furrows, short-cycle crops such as beans, corn, tomatoes, beans, etc. can be planted, which generate income in a short time, allowing the farmer a continuous cash flow[15], [16], [17], [18], and [19].

Good Agricultural Practices. Include a set of activities aimed at improving traditional methods of agricultural production, taking as main parameter the safety of the product, without causing great impact on the environmental component in relation to flora, fauna, and the welfare of workers. Its application is oriented to the production of fresh, healthy agricultural products of excellent quality, and that generate greater economic income for the farmer. From the input of the participants it is advisable to increase the use of organic fertilizers and inputs, and only resort to the use of synthetic agrochemicals in cases where their application is strictly necessary. In the case of pesticides, their application must be restricted, their use will be justified and rational only for those cases where the threshold of economic damage so requires. [20], [21].

Cocoa pruning. Is a very important practice in the management of the cocoa crop, because its proper execution manifests itself in the increase of productivity, but poorly carried out also has its effects as deterioration of the plantation, little formation and maturation of ears, and consequently decrease of income. The pruning seeks to modify the structure of the tree allowing it to maintain a balanced shape in terms of its general architecture, that is, a crown must be maintained with a height not greater than 3 meters, and the lateral branches are pointed out to avoid intersections. This work of pruning allows a better entry of sunlight that helps the maturation of the ears, greater aeration, and induces a decrease in the appearance of diseases and appearance of pests. By pruning encourages the influence of the factors of productivity, leading to the presence of a large quantity of logs on which to form the cushions, increases the fertilization of the flowers, promotes the formation of new leaves expand the leaf area, and in general contributes to obtaining good crops. All the material coming from the pruning and the thick branches are repelled and left on the ground as cover [22], [23]. The types of pruning are.

Formation: it is done between 12 and 24 months after sowing, in which the lateral suckers are cut, trying to leave an erect and balanced stem in the structure of the tree.

Maintenance: this activity is done throughout the life of the tree, and is done in the dry season.

Sanitary: it serves to remove sick parts of the tree, fruits and branches.

Rehabilitation: it is done in crops with low production. Pruning time. When there is a low fruit load on the tree. Finishing the harvest peaks. At the beginning of the rainy period. When there is low flowering. Before beginning the vegetative period. Fertilization. Is an important practice for the good development of cocoa trees, and can be done in an organic and/or inorganic way, according to the nutrient deficiencies that the soil analysis yields. A record should be kept of the results of the analysis of soil, nutritional need of the crop, soil fertility.

Management of by-products. Of all the harvested cobs, those that present some symptom of disease must be separated, in some cases this separation process can reach up to approximately 40% of the total harvested cobs, which are thrown to the floor along with the shells of the newly split cobs. The shells represent approximately 79%, and are placed stacked between the streets of the cocoa crop to begin the decomposition process, to later use it as an ingredient for the preparation of compost. Diseased ears and freshly broken cocoa shells are treated with a cupric fungicide and then covered with lime, in order to reduce their influence as a source of inoculation of diseases such as Monilia. Once the lime is applied, the shell stack remains inside the batch in a decomposition phase for approximately 20 days, and can then be used for compost preparation.

Compost. It is the process of transformation of some elements present in the materials we use as organic fertilizers. In this process microorganisms that could contaminate plants or soil are inactivated, all activity is aerobic, therefore, it is done in the presence of air, and the decomposition of matter is performed by fungi and bacteria. Composting should be done in a place near or inside the cocoa crop, where we find the materials that are going to decompose. Once the site for composting is defined, preferably a very simple house covered with plastic should be built to avoid contact with rain, around this structure a drainage trench is built to avoid waterlogging.

3.3 Conditions of Cocoa processing to obtain quality almonds.

The activity of profit is the most important stage of the cycle of cultivation of cocoa, which starts with the collection process only cobs, mature, separating that show signs of disease that you have reached the point of maturity appropriate, then, within the same culture it comes to the game of the same, and finally comes to the shattering, trying to draw only the grains without it being attached to part of the placenta. The cocoa profit includes those operations that are made to the bean after the harvest. However, this manual includes harvesting operations, as these have a great impact on the final result of the quality of almonds or grains. With this consideration, the benefit includes harvesting and shelling, fermentation, drying, sorting, sorting, packing and storage operations, which will be dealt with below [24], [25], and [26].

Harvesting of mature ears. The frequency of harvesting depends on the health condition of the cobs, however, it is preferable to avoid very long intervals between harvesting, as there is a risk of harvesting cobs with different ripening states. In any case, the ripe cobs should be discarded, as they may contain grains in the process of germination in which during the fermentation process or the drying process the embryo can be detached, allowing the entry of molds or insects. Unripe ears should also not be used, as in this case the contained grains are difficult to separate and remove during the sorting and cleaning processes due to their compaction state. The ideal state to harvest the ears is when they are ripe. However, at the time of harvesting, not all are in that state, and the ears that have just begun to ripen are also collected. In the peak periods of harvest, harvesting rounds should be conducted weekly. However, in low-yield seasons, harvests can be scheduled every two to three weeks.

It is essential not to leave the ears to ripen because they can be contaminated with some diseases with fungi. This state favors the germination of the grains, which is considered a quality defect. For no reason should fruits still be harvested green. Grains from green ears are hard, cannot be easily separated and do not ferment because the mucilage has not yet formed. The harvest of the cobs must be carried out with appropriate techniques and tools. Generally the ears are cut with pruning shears. However, those that are not within reach are cut with sharp hooks or half-moons attached to a long handle of light wood. The cut is made without damaging the branch from which the cob is seized. When the presence of the disease known as "mal del machete" or "macana sore" is suspected, it is advisable to disinfect the harvest tools every day that this operation is going to be performed [24], [25], and [26].

Departure of the ears. The departure of the ripe cobs, preferably should be done by a blow with a wooden mallet instead of using the machete system, because this practice can lead to the operator is caused cuts, or that on the surface of some cocoa almonds there are wounds becoming gateway to mold and insects inside the grains. Before splitting the ears, it is important to separate the healthy from the sick, with insect or animal damage. This is the starting point to avoid the deterioration of the quality because only the recovered healthy grain can be mixed with the grain from healthy ears. Suspicious grain of poor quality is processed separately. It is important to insist that for no reason should healthy grains be mixed with grains whose external appearance indicates damage from pests or diseases, since this would deteriorate the final quality of the product. Usually, the most used tool to break the ears is the machete. However, this tool has several drawbacks: the risk to the operator, the possibility of cutting the grains and the low yield. In some regions a very simple device is employed that consists of a machete embedded by the sharp part to a piece of vertical board properly supported on a piece of horizontal board The ears are usually opened in half hitting them on the back of the machete, holding the half with one hand and exerting a slight twist on the other half. The advantages of this device are based on the safety for the operator, the speed of operation and the reduction of damage to the grains [24], [25], [26].

Shelling of the ears. Once the ears are split, the seeds of the shells are extracted, sliding the fingers of the hand along the central vein (also called placenta), taking care not to extract it, if it is detached it must be separated at once, because if left in the grain mass it becomes impurity that ultimately goes to the detriment of the quality of the product. The grains are extracted with the fingers leaving the placenta attached to the cob and pieces of bark, leaves, etc., mixed with the grains are removed. For this operation it is customary to use a glove that prevents tearing of the nails. In the places where the ears are to be shelled it is advisable to place a plastic on the surface of the soil to prevent the grains from mixing with soil, leaves or grime in general. It is advisable to shelling in separate containers the healthy grains and those of altered appearance which must be fermented and dried separately. After opening the cobs, the beans should be fermented before 16 17 Cocoa benefit for 24 hours. For no reason can grains from open ears be mixed on different days [24], [25], and [26].

Fermentation. It is the main step of the cocoa benefit, because it is in which the flavor and aroma of the product is developed. It is a process in which a series of biochemical and physical changes occur on the cocoa bean. This activity must be carried out in an enclosed place, to avoid drafts that may influence the normal development of the fermentation process. It is the process that continues after the shelling. It consists of stacking the grains for several days so that the microorganisms break down the mucilage (the white and sugary pulp that surrounds the grains), increase the temperature to produce the death of the germ or embryo and initiate biochemical changes and enzymatic reactions inside the almonds, which will be responsible for the formation of the precursor compounds of the chocolate flavor, This process also facilitates the drying of the grains [24], [25], [26].

Fermentation is carried out by a succession of microorganisms (yeasts, lactic acid bacteria and acetic acid bacteria), which begins in anaerobic conditions (without the presence of oxygen) and ends in aerobic conditions (in the presence of oxygen). When the pulp decomposes, there is heat release and the temperature of the dough can reach 45-50ºC. Alcohol, lactic acid and acetic acid are formed that permeabilize the head or membrane that covers the grains. These compounds and the temperature, participate in the death of the embryo, propitiating the dissolution and diffusion of pigments (anthocyanins) and alkaloids (theobromine and caffeine), which produce the bitter taste of the finished product. The death of the embryo is indispensable for the biochemical processes that take place within the grain to be triggered [13], [25], and [26].

Drying process. Finished the fermentation process of cocoa beans, it is necessary to subject it to the process of drying, which consists of reducing the moisture that comes out the cocoa bean once the stage of fermentation (55 per cent) to a moisture content of final maximum 7%1 which is the accepted during the commercialization phase of the cocoa bean dry, which allows to preserve the quality of the grain during storage and without risk of deterioration by fungi. In the cocoa beneficiation manual after fermentation the cocoa must be dried immediately, not only to remove the moisture from the bean that must remain at 7%, but also, so that some biochemical reactions that will finally produce the precursors of the flavor continue. Drying is as important as good fermentation. The moisture content of dry grains should not be higher than 8% due to the propensity of the grains to mold, nor should it be below 6%, because the grains become brittle and brittle [13], [25.

Natural drying. There are two ways and the first is natural drying which is usually done in the sun on wooden platforms. For the process to be uniform, the first day the grains must be stirred infrequently and in the following days more often until the process is finished. The best sign that drying is over is the cracking or crunch felt when squeezing a handful of the grains in the early hours of the morning. At the end of drying, inside the beans develops kidney shape structure and the brown color typical of well-benefited cocoa. Depending on the hours of sun and the intensity of the sun's rays, drying usually takes between 3 and 9 days. The drying operation is used to remove flat grains, wrinkled, black, moldy, small, stuck, and damaged by insects, broken and all foreign material found. The drying area is calculated taking into account the dry cocoa production of the farm in the peak week for which the production is multiplied by 0.1 which is the conversion factor of 2 meters per kilogram [13], [24], [26]..

Artificial drying. The second way to perform drying is in some regions, the harvest season coincides with the winter season and under natural conditions it is difficult to dry the cocoa due to the low luminosity. In this case, an artificial drying system must be used because otherwise the drying in the sun would be very prolonged and the grains would run the risk of becoming contaminated with fungi, deteriorating their quality. Hot air is used for this type of drying. The simplest artificial dryer is basically a permeable platform (wood, metal or perforated aluminum sheet) horizontally recessed over a chamber provided with a coal or fuel burner coupled to a metal tube with its rear end attached to a chimney whose diameter must be at least 10% of the diameter of the tube. In the perforated sheet the gaps should be 3 centimeters apart and their diameter or length should not exceed 1 centimeter. Hot air comes out of the platform and dries the cocoa beans. As the back end of the platform heats more than the side of the burner, it is necessary to flip the grain mass gradually exchanging its position so that the drying is as uniform as possible [13].

To facilitate drying, the thickness of the mass of grains should be about 12 centimeters, the temperature of the hot air cannot be greater than 60 ºC in order to avoid the inactivation of fundamental and indispensable enzymatic reactions in the formation of flavor precursors. High temperatures anticipate the process of roasting or roasting the grains, which become brittle and brittle. With a temperature of 55 – 60 ºC the drying process is completed in 30-34 hours.

Selection. After drying, the grains must be subjected to a thorough cleaning eliminating all foreign materials mixed with them (pieces of bark, placenta, etc.), black grains, moldy, damaged by insects, broken, wrinkled, stuck, pieces of husk and dust. Normally sieves are used to make the separation and fans to blow the dust and pieces of husk.

Classification. The frequency of flat and very small grains (less than 1 gram per grain) in a cocoa batch varies according to planting material and growing conditions. An examination of a sample will tell whether sieves are needed to separate these grains. When necessary, the grain is passed through sifters or specific sieves to separate the first quality cocoa from the "pasilla " (grains with less than 50% of almonds, for this effect the sifter is used No. 6 which is special for cocoa. Premium cocoa is composed of whole grains with a large size margin.

Packaging and storage on farm. After the classification, the packaging is carried out, the dry, clean and classified grain is packed in bags or sacks of fique or jute new or in perfect condition, which have not been used before to pack products with penetrating odors (fish, for example). After packing the grain is weighed in such a way that the net weight of a bag is 62.5 kilograms or 16 bags per metric ton. The storage of cocoa beans on farms has serious drawbacks. Adequate and safe holds are generally not available, and when grains have not dried properly, they are prone to mold invasion and insect infestation. In the tropics cocoa supports storage periods of 2 to 3 months without assuming risks in its quality. It is best to market the grain immediately after its packaging in local markets or in the direct agencies of the manufacturers.

3.4 Factors of physical and sensory quality for the Commercialization of cocoa almonds.

Taste is a key quality factor for products made from cocoa. It includes the intensity of the cocoa or chocolate flavor, the absence of unpleasant flavors, and presence of secondary aromatic notes. Defects include incomplete fermentation, over-fermentation and contamination. The cutting test, used for the assessment of fermentation quality, provides an indication of organoleptic defects such as excessive bitterness and astringency due to the presence of slate grains, or perhaps a mildew/moisture taste product of moldy or infested grains. However, the cutting test is not a reliable indicator of taste quality [25], [26].

To assess the taste of a sample of cocoa beans, the sample must first be converted into liquor or cocoa paste, or chocolate. Then, we proceed to taste by a tasting panel formed by a group of five to ten experienced tasters. In some parts a single expert taster evaluates both for the determination of unwanted flavors and detailed description of aromas, always additional tastings are carried out to ensure greater statistical rigor. The evaluation of the liquor is more demanding, but it can be tasted directly, without the need to make mixtures with cocoa butter, sugar, milk or other products that mask the true flavor and provide flavors and aromatics not present in the natural state of the cocoa bean used in the tasting.

Chocolate also requires a period of time for the taste to stabilize after processing; it is more difficult to preserve than liquor, both frozen and at room temperature, and it is often difficult to make chocolate that meets the usual standards on farms and farms where the grains are produced, due to the lack of processing equipment.

Unwanted mold flavors. During drying or storage the use of firewood or other sources of heat generation may occur that can produce fumes that at the same time cause contamination of the cocoa almond by characteristic flavors of smoke in the liquor and later in the chocolate. Smoke flavor is another unwanted flavor that cannot be removed during chocolate production. Smoked cocoa beans can be easily detected in a sample by crushing or grinding some beans and smelling them.

Unwanted smoke flavors. The unwanted taste of smoke is sometimes compared to the taste of Ham, because it is reminiscent of smoked ham. These flavors can also be generated by exceeding the fermentation time, but it is quite easy to differentiate one from the other. The unwanted taste of smoke in contaminated grains, the taste of Ham prevails. While in the over-fermented grain, the notes of smoke are minor and the putrid, ammoniacal, sometimes soapy or phenolic background prevails. Phenols such as guaiac are thought to be present in smoke-contaminated grains and in over-fermented grains. To avoid the above, it is necessary to adjust the variables to be controlled in fermentation, especially the time according to the variety of cocoa and prevent the smoke effect in drying and storage [24], [25], [26].

Unwanted flavors to earth. They occur in cocoa almonds located or located on the outside of containers, sacks or silos where they are stored by the accumulation of dirt, dust or soil residues or other waste. Sometimes they manifest in a very unpleasant way, when they are presented in significant quantities the decision is made to reject the lot [24], [25], and [26].

Acid taste. It is produced by the presence of excessive amounts of acidic compounds that are generated in fermentation. Two acids are linked: acetic acid (volatile) and lactic acid (non-volatile). A proper drying process can reduce acidity in fermented grains, but when dried too quickly, acidity persists. During the transformation of the cocoa almond due to the effect of the rise in temperature, acetic acid decreases to low acceptable levels, but lactic acid, because it is not volatile, remains present; the lactic acid present in excess generates an unwanted flavor in chocolate products. At the same time, the excess of lactic acid present prevents the development in its splendor of the taste of chocolate [24], [25], [26].

Bitterness and astringency. The complex taste characteristics of chocolate include some attributes of bitterness and astringency, but when presented excessively it is considered unpleasant. Likewise, like other unwanted flavors, their undesirable characteristics cannot be reduced or eliminated by activities during normal industrial processing. These flavors are associated with insufficient fermentation and / or certain varieties. Cocoa almonds above 3% of grain content unfermented or slate, produces very bitter and astringent liquors, produce in chocolate characteristics of excessive astringency. In that order, completely purple cocoa almonds also generate bitter and astringent flavors. So it is necessary to avoid excessive amounts of slate and purple grains. A cocoa processed with good manufacturing practices harvested, fermented and dried in the sun contains a high percentage of partially brown/purple beans [24], [25], [26].

3.5 Physical characteristics.

Uniformity. For the marketing of cocoa almonds, it is necessary to ensure the uniform and homogeneous quality of cocoa, between different bags within the same batch, as well as in different batches from the same supplier or brand. This makes manufacturers give greater value to the reliable supply of cocoa beans of uniform quality, since it allows them to manufacture excellent quality and uniformity chocolate. Uniformity includes the number of defective grains, the size of the grains and the degree of fermentation. Sometimes uniformity is achieved by mixing beans, but all cocoa used in the mix must have the same classification. It is not recommended to mix low quality cocoa with excellent quality cocoa to obtain a cocoa that only meets minimum quality cocoa specifications [24], [25].

Edible material yield. It is considered one of the factors that affects the price of cocoa willing to pay by the chocolate manufacturer because it directly influences its performance. In this, several factors are linked that quantitatively affect the edible material, that is, the grain without the shell and the yield in cocoa butter to be obtained from a certain batch of cocoa. Among these factors are grain size, shell content and fat content. Depend on climatic and genetic factors; some factors can be improved by the application of good postharvest and transport practices. The cocoa producer can improve the quality of cocoa by removing foreign matter while processing in its production unit, foreign matter such as stones, wood residues or residues of the cocoa itself impair the quality of the almond. A batch of cocoa without foreign matter is valued at a better price in favor of the cocoa producer [24], [25].

Wet content. The moisture content of cocoa almond should be about 7%. If it is greater than 8%, it means a loss in the yield of edible material, as well as an increased risk of growth of contaminating microorganisms such as molds and bacteria, which goes against food safety and safety, sensory characteristics and quality for its transformation. In this sense, if the moisture content is below 6.5%, the shell is too fragile and the grains break easily, which causes its fragmentation into small particles with a high proportion of broken grains, causing the activation of higher levels of lipolysis and greater generation of free fatty acids[24], [25].

4. Conclusion

In the present research, from the economic point of view, cocoa cultivation has an important participation in the northern coast of the country. It was verified the effort made by small producers to integrate in the management of their cocoa crops, the processes, harvesting and post-harvest activities that are carried out to obtain dried cocoa almonds for marketing. However, technical assistance is needed to improve actions, activities and processes in small production units. Likewise, it was proven that cocoa is the basis for the support of many families of agricultural populations in the rest of the country, with a vocation for agro-industrial work similar to San Jose de Oriente and Media Luna in the Department of Cesar. On the other hand, it is necessary to link trade and territorial entities to the implementation of strategies that contribute to the improvement of cocoa production for the export of almonds or cocoa beans rich in flavor and aroma or also called Special Cocoa worldwide. In addition, it is important to mention the support of the Universidad Popular del Cesar and the Agroindustrial Engineering program

References

[1] "Cocoa". Editorial team, Etc. Argentina. Available at: https://concepto.de/cacao /. Last edited: August 5, 2021. Accessed: August 26, 2021 - Source: https://concepto.de/cacao/

[2] National Cocoa Farmers Federation. (2020). PRESS RELEASE - This is the ranking of cocoa production in Colombia. http://www.fedecacao.com.co/portal/index.php/es/2015-04-23-20-00-33/1193-boletin-de-prensa-asi-quedo-el-ranking-de-produccion-de-cacao-en-colombia

[3] Minagriculture. (2020). COCOA CHAIN Management of Agricultural and Forestry Chains. https://sioc.minagricultura.gov.co/Cacao/Documentos/2020-03-31 Sectoral Figures.PDF file

[4] National Cocoa Farmers Federation. (2019). ICCO ratifies Colombia as an exporter of fine cocoa and aroma. https://www.fedecacao.com.co/portal/index.php/es/2015-04-23-20-00-33/867-icco-ratifica-a-colombia-como-pais-exportador-de-cacao-fino-y-de-aroma

[5] National Cocoa Farmers Federation. (2020). PRESS RELEASE - This is the ranking of cocoa production in Colombia. http://www.fedecacao.com.co/portal/index.php/es/2015-04-23-20-00-33/1193-boletin-de-prensa-asi-quedo-el-ranking-de-produccion-de-cacao-en-colombia

[6] National Cocoa Farmers Federation. (2021). Legal Framework. http://www.fedecacao.com.co/portal/index.php/es/fondo-nacional-del-cacao-4/marco-juridico

[7] De la Rosa, L. M. (2019). Emerging paradigms in management. Mexico, Mexico: Pearson Educación de México, S.A. de C.V. Retrieved on July 28, 2021, from http://cathi.uacj.mx/bitstream/handle/20.500.11961/8253/EBK%20De%20la%20Rosa%20Paradigmas%20emergentes%20en%20la%20administracio%CC%81n%2001%20SE%20HA%2049652.pdf?sequence=1&isAllowed=y

[8] EYSSAUTIER DE LA MORA, M. (2002). Research methodology. Development of intelligence. Mexico: ECAFSA.

[9] HERNANDEZ SAMPIERI, R., Fernandez Collado, C., & Baptista lLucio, P. (2010). Research methodology. Mexico: Mc Graw Hill.

[10] HURTADO DE BARRERA, J. (2010). Research methodology. Bogota - Caracas: SYPAL - CIEA. Quiron editions.

[11] C.R. Vidal Tovar, C.A. Cadavid Osorio, G.E. Angulo Blanquicet and C.A. Severiche-Sierra. Obtaining a sugar syrup from the use of the extract of the guácimo fruit (Guazuma ulmifolia lam). IOP Conference Series: Materials Science and Engineering. 2020. Tomado de: https://doi.org/10.1088/1757-899X/844/1/012003

[12] Prieto, D. (2019). TECHNOLOGY AS A TOOL TO INCREASE AGRICULTURAL PRODUCTIVITY IN COLOMBIA. https://repository.uamerica.edu.co/bitstream/20.500.11839/7246/1/2142125-2019-1-EF.pdf

[13] Cubillos, G., Jaime, G., & Correa, E. (2008, October). COCOA BENEFIT MANUAL 2008. For: technicians, professionals in the agricultural sector and producers. 17. https://chocolates.com.co/wp-content/uploads/2020/06/manual_beneficio_cacao.pdf

[14] Contreras, C. (2017). Analysis of the cocoa value chain in Colombia: generation of technological strategies in harvest and post-harvest operations, organizational, installed capacity and market [National University of Colombia]. https://repositorio.unal.edu.co/bitstream/handle/unal/60801/10323734482017.pdf?sequence=1&isAllowed=y

[15] Guerrón, R. (2018). Feasibility project for the production of cocoa with a view to export in the farm "Lesly" located in the Province of Santo Domingo de los Tsáchilas. http://repositorio.ucsg.edu.ec/bitstream/3317/11750/1/T-UCSG-POS-MFEE-135.pdf

[16] Lopez, A. (2019). PROPOSAL FOR THE CREATION OF A CONSORTIUM AIMED AT THE EXPORT OF COCOA PASTE TO THE REPUBLIC OF ARGENTINA. https://repositorio.uide.edu.ec/bitstream/37000/3084/1/T-UIDE-1126.pdf

[17] Ramos, A., Gomez, M., Machado, E., & Aranguren, Y. (2020, September). Phenotypic and genotypic characterization of cacao cultivars (Theobroma cacao L.) from Dibulla, La Guajira, Colombia. http://revistacta.agrosavia.co/index.php/revista/article/view/1557/774

[18] Ramos, D. (2014). Environmental considerations and socio-economic viability of the cocoa production system. The case of the farmers of Pueblo Bello, Department of César. [PONTIFICIA UNIVERSIDAD JAVERIANA]. https://repository.javeriana.edu.co/bitstream/handle/10554/15004/RamosPerezDaniloFernando2014.pdf?sequence=3&isAllowed=y

[19] SENA, CAM, FAM, ADAM, & Coffee Company Huila Ltda. (2008). Agreement of concertation for a cleaner production in the coffee subsector - Special coffee,. Second Ed, 56. https://www.cam.gov.co/images/documents/phocadownload/produccion_sostenible/Guia_cacaocultores.pdf

[20] García, H., & Martínez, R. (2018). DETERMINATION OF THE FLAVOR PROFILE OF TWELVE NATIVE CACAO TREES (Theobroma cacao L.) PRODUCED IN SEVEN COCOA FARMS IN EL SALVADOR. [UNIVERSITY OF EL SALVADOR]. http://ri.ues.edu.sv/id/eprint/16512/1/13191670.pdf

[21] Vargas, E. (2018). THE PRODUCTION AND EXPORTS OF COLOMBIAN COCOA BETWEEN 2007 AND 2016: CHALLENGES TO ACHIEVE GREATER COMPETITIVENESS IN THE INTERNATIONAL MARKET. https://repository.uamerica.edu.co/bitstream/20.500.11839/6929/1/5131561-2018-II-NIIE.pdf

[22] SHIN, C. (2017). Analysis of the components and approaches of the Kwakumey Ati project for the sustainable development of the Sierra Nevada de Santa Marta. https://repository.javeriana.edu.co/bitstream/handle/10554/21853/ShinChoi2017.pdf?sequence=1&isAllowed=y

[23] International Cocoa Organization. (2015). Analysis of the international Fine and Aroma Cocoa market: Opportunities for Colombia. https://www.swisscontact.org/fileadmin/user_upload/COUNTRIES/Colombia/Documents/2__Analisis_de_Mercado-Carlos_Contreras-Corpoica.pdf

[24] CAOBISCO/ECA/FCC Cocoa Beans: Chocolate and Cocoa Industry Quality Requirements. September 2015 (End, M.J. and Dand, R., Editors).

[25] Horta-Téllez, Heidi Briggity, Sandoval-Aldana, Angelica Piedad, Garcia-Muñoz, Maria Cristina, & Cerón-Salazar, Ivonne Ximena. (2019). Evaluation of the fermentation process and final quality of five cacao clones from the department of Huila, Colombia. DYNA, 86(210), 233-239. https://doi.org/10.15446/dyna.v86n210.75814

MARQUINEZ MONCAYO ANGELO ABIGAIL. (2021). Agronomic and economic efficiency of compound fertilizers in cocoa CCN-51. Degree work report. Prior to obtaining the title of agricultural engineer. Modality: research project. Manuel Félix López Polytechnic Agricultural School Of Manabí. Calceta, Ecuador. Taken from http://repositorio.espam.edu.ec/handle/42000/1433

Use of augmented reality technologies in the framework of m-Learning at the Infantery Museum Marine Park

F. Oñate[1], E. Vergara[1], A. Morales[1], S. Sanchez[1] and J. Barreto[2]

[1]Antonio José de Sucre University Corporation (Colombia).

[2] Infantery Museum Marine Park

Docente_investigador6@corposucre.edu.co

Abstract. The research proposes to design a software to visualize the relevant information of historical military objects through augmented reality technology, contributing to the modernization of the museum, generating a more attractive and autonomous visit for visitors. The research is embodied in two phases. In the requirements phase, data collection instruments were implemented in order to identify the functional and non-functional requirements of the proposed system; In the design phase, the block diagram and flow diagram are modeled to outline the software architecture and illustrate the communication between the different technologies that make up the system. As a result, the structure and characteristics of a model that meets the needs of the Infantery Museum Marine Park was obtained. It is concluded that this model can contribute to the digitization, exhibition and appropriation of the cultural content exhibited in the different scenarios of the Infantery Museum Marine Park in its processes of transmission of knowledge about the various cultural heritages of the museum.

1. Introduction

With Decree 457 issued by the National Government of Colombia and signed by the President of the Republic in March 2020, Mandatory Preventive Isolation was established "for all the inhabitants of the Republic", [1] And, since these prevention measures began to be implemented, among other restriction measures taken by most of the world's governments due to the Covid-19 pandemic, many public and private establishments have been obliged to cease their work activities with clients, that is, they had to maintain their physical facilities without access to the general public due to the restrictions regarding physical contact in closed and public establishments, dictated and imposed by the National Government of Colombia. Within the definition of establishments that had to close their doors, are the museums - in which the Infantery Museum Marine Park is located located in the department of Sucre (Colombia) - these are in charge of incorporating, maintaining, investigating, disseminate and exhibit all kinds of content about the heritage, both historical and cultural, of civilizations. A UNESCO analysis of different reports shows that in 2020 almost 90% of the world's museums (approximately more than 85,000) have had to temporarily or permanently close their doors to the population.[2]. In the Museums and Heritage edition of the periodic magazine Culture and Development, the thought of Herman Van Hooff, director of UNESCO's Regional Office for Culture for Latin America and the Caribbean, stands out, highlighting the importance of museums and cultural centers as spaces full of opportunities for the transmission of cultural values, knowledge and identities, and their contribution to the regional economy and job creation[3]. On the other hand, emphasizing one of the goals of the Sustainable Development Goal in the Education 2030 Framework for Action, exactly # 4, which is related to education and especially with inclusion and non-discrimination in access to the foundations of teaching, which includes allowing free access to knowledge through museums. Thus, like all the inconveniences that museums have had to be able to operate and provide their services properly, due to the pandemic, these have been the trigger for museums to consider the use of industry 4.0 tools and innovative strategies for the sake of digitization and generation of new experiences during the exhibition and promotion of its historical and cultural collections.[4]Some museums have integrated data mining into their schemes to improve the data-driven decision-making process. On the other hand, in order to reduce the risk of contagion among visitors, the Livingstone Museum took the measure of closing its physical doors, but continues to operate actively through Facebook and its web portal.[5].

For this reason, the objective of this research is to design a software prototype that makes use of augmented reality through marker-based monitoring. In this way, you can view objects and digital information related to the cultural content exhibited in the Infantery Museum Marine Park.

The document is organized as follows: initially the concepts of Augmented Reality and the development tools or kits, necessary to implement this new technology to any project, most used to date, are highlighted, then some of the The importance of Augmented Reality in the context of museums, the methodology used in this study is subsequently described and illustrated with tables and images, the technical user requirements and block diagrams, flow diagrams, the system architecture and finally the conclusions of the investigation.

2. Theoretical framework

2.1. Augmented Reality (AR)

It can be represented as the integration of digital content (images, text, 3D modeling, videos) to real environments in order to enrich its perception in a non-immersive environment, therefore in the following section the main concepts and theories that form fundamental part of the present investigative work. According to Innovae, a company that is dedicated to working or developing projects related to virtual reality and augmented reality with more than 16 years of experience in the context of industry 4.0, this company defines that augmented reality integrates graphic content on a point in the plane of real life. Use smart mobile devices and / or glasses, which add certain digital information to the reality that the user sees, If smart mobile devices (such as phones and tablets) are used, the information is captured through the device's camera. On the other hand, if augmented reality glasses are used, the result is shown in the form of a projection on the lens of the glasses that generates a hologram sensation. Taking the previous definitions, it can be argued that Augmented Reality is a technology that allows digital interaction with the real-life environment in order to provide a semi-realistic and interactive experience for the user.[6].

To develop solutions based on augmented reality, it is necessary to use, depending on the purpose of the project, some technologies and development tools or kits that facilitate the integration of augmented reality into the project, among these tools are:

2.1.1. Vuforia Engine

According to the Vuforia Engine Developer Libray site, Vuforia Engine is the most used platform for the development of Augmented Reality and Mixed Reality applications, compatible with mobile phones, tablets and augmented reality glasses. Allows developers to add advanced computer vision capabilities to their installable files for Android, IOS, and Universal Windows Platform (UWP) applications to create realistic experiences with real-world objects[7].

2.1.2. Unity
It is a framework for the development of applications that integrate 3D and 2D content in real time, it is one of the video game engines and a powerful multiplatform Integrated Development Environment (IDE) most used by designers and developers. in favor of creating interactive and immersive experiences [8].

2.1.3. Flutter
It is a package of portable user interface tools that belongs to Google, and that is also open source and free, that is why it is used by companies and developers around the world such as BMW, NU Bank, Ebay, among others. , this allows you to create native and efficient compiled applications for mobile devices, web and desktop operating systems starting from a single existing code base [9].

2.1.4. ARCore.
Google owns ARCore, a platform that allows you to create augmented reality experiences. This can be used with different APIs so that the mobile device (smartphone or tablet) detects its environment, understands the world and interacts with the data it extracts from it. To enable all the integration between virtual content and the real world, ARCore uses three key elements, which are: motion tracking, environmental understanding and light estimation. In this way, ARCore can detect the location of the mobile device while it is moving and autonomously develop a mapping of the real world.[10].

2.2. Follow-up marker-based
In the framework of Augmented Reality and Mixed Reality, markers are a series of images or objects linked to the application, these take the role of triggers of virtual information in it. Marker-based tracking is capable of using a range of different kinds of markers or targets, these include QR codes, reflective physical markers, imaging targets, and 2D labels, objects can be cubic, cylindrical in shape or a 3D layout of any real object [11].

2.2.1. Image objectives
They are a specific type of marker that the user registers manually in the application and they are responsible for showing the desired virtual content, these objects are recommended to have high contrast [11], a size of 2.25MB or less[12], as well as diffuse, uniform and reasonably bright lighting to facilitate the process of detecting and monitoring these. In addition, the image objectives must comply with a series of physical properties for the operation of the application, among which are: the image object must be on a rigid surface, such as cardboard, or paper on a non-flexible surface, the size of this may vary depending on the type of real object or the distance at which the image objective is going to be located, for a longer duration of the printing quality, the material on which it is printed should not bend, wrinkle or roll up, also this material should not reflect the ambient light intensely, this could make it difficult to track the target image at certain angles[13].

2.2.2. Objective models
These consist of 3D models of real life objects, such as appliances, vehicles or everyday objects such as cups or toys, these can be 3D scanned by third party applications or they can be 3D CAD models, it is recommended that these 3D models, not are flexible and do not have shiny surfaces [14].

2.2.3. Objective Objective
These consist of real life objects, these are not usually large such as toys, action figures, among others. Using this method for object recognition provides the user with a more interactive, immersive, and enriching experience. To implement this type of markers, it is recommended that the object does not reflect the ambient light (which should not be opaque and not very bright) in an intense way, that it does not contain shadows from other objects, that it is not flexible or with moving parts, the surface

of the object must have texture and a good contrast, in addition to these characteristics, Vuforia Target Manager only allows .OD (object data) files created by the application scanner for the generation of object targets [15].

2.3. Museums and Augmented Reality

In the book Interculturality of Ethnicities in Colombia, the importance of the incorporation of new technologies is highlighted, not only for the preservation of cultural heritage but also to give added value to cultural heritage, this in order to allow the user to enjoy of more interactive and enriching experiences, as an example of this the case of the Mataró Museum stands out, which implemented an augmented reality system with the exhibition "Mar de Fondo", which aimed to improve the experience of visitors with the works exhibited in the museum, this system allowed the user to point with the camera of a smartphone or a Tablet to the work of interest, in this way the application showed the relevant digital content through pictures about the work [16]. Regarding the Infantery Museum Marine Park, it opened its doors to the public on August 17, 2012 in order to exhibit the history of the Colombian Marine Corps.[17]. The museum offers information about the Colombian Marine Corps and has recreational spaces so that the process of appropriation of information, of the different stages of the Marine Corps, is as clear as possible.

2.3.1. Immersive experiences in showrooms

In the article Multimedia guides with augmented reality in museums of the 21st century: virtuality as an integral part of the exhibition discourse, the great work that Augmented Reality has in spaces where all kinds of works or cultural objects are exhibited is highlighted because this provides a range of possibilities that would not normally exist, thus increasing the degree of public capture due to the novelty of the use of new technologies that provide the visitor with an unusual experience[18].

3. Methodology

This research fits into the definition of two types of research scope, the first is the definition of basic research of descriptive scope, which, according to the book Research Methodology. Sixth edition, this type of study seeks to measure or collect information on the concepts, variables, groups, communities, objects or people subjected to analysis without influencing their behavior and the data collected and, in addition[19], "Consists of experimental or theoretical work that is undertaken mainly to obtain new knowledge about the foundations of observable phenomena and facts, without having as objective any specific application or use, regardless of the area of knowledge" [20].

The second is the definition of applied scientific research, in this type of research "the development of prototypes is not obtained as results, however, within the activities that make up scientific research, prototypes could be developed to validate a hypothesis, but in non-environments. relevant " [20].

Within the research process, an analysis of the state of the art was carried out to learn about the latest advances in the application of Augmented Reality in museums and to determine the minimum technical requirements and technologies that the system to be designed must have.

The final purpose of the research is to generate the design of a mobile application that allows viewing through augmented reality technology the relevant information on historical military objects in order to contribute to the modernization of the museum and make a visit more attractive and autonomous for visitors.

The research is embodied in two phases, taking into account the software development methodology called Waterfall.

In the requirements phase, the documentary analysis of the information organized with the bibliographic matrix and analyzed with the content analytical matrix was carried out, in addition to the application of semi-structured interviews to personnel of the museum of the marine infantry, in order to to identify the functional and non-functional requirements of the system proposed in the present investigation.

For the design phase, block diagrams and flow diagrams were used to outline the architecture of the system and illustrate the communication between the different technologies that make up the system.

4. Results

Through semi-interviews applied to the administrative staff of the Marine Infantry Museum Tour and the documentary analysis and with the use of the bibliographic matrix and content analytical matrix instruments, a survey of requirements was carried out (see table 1), based on these requirements, the proposal of the system architecture is proposed, by means of a diagram of the operation of the software and flow diagrams.

Table 1. System requirements

Functional requirements	Non-functional requirements
The application must suggest a route of movement through the museum.	The Graphical Interface must be intuitive so that the user has a correct idea of the use of the application.
The System must superimpose digital content with the use of the camera, such as cultural objects and expressions in multimedia format.	The application must sort the museum objects by categories
The system must recognize image, 3D model or object target type markers.	The system must meet the standards of usability and portability.
The application, in the main module, should allow the user to choose the consequent action by means of choice buttons	For the development of the application it will be necessary to use Unity 3D as a graphical development environment to which a specific library for augmented reality called Vuforia is incorporated.
The application must show images of the elements of the museum.	For the development of the graphical environment of the application, it will be necessary to use Flutter for the design of widgets and functions of the application windows.
The application must show pertinent information about the cultural objects exhibited in the museum.	
The system must be able to recognize under normal environmental conditions, the different targets within the museum.	
The application should have a suggestions section for the respective feedback of its functions.	

After the analysis of the requirements where the limits and risks of software development are determined, we proceed to the design of the architecture of the proposed prototype. The results of this design are shown below:

4.1. Software Operation Mapping

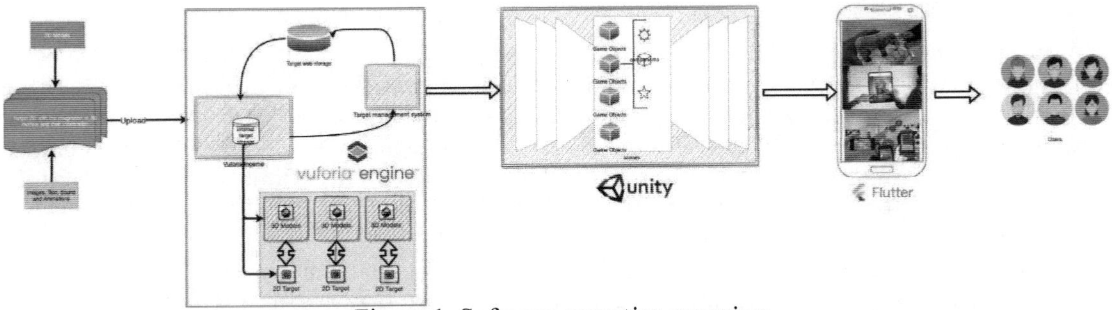

Figure 1. Software operation mapping.

The operation of the software consists of the use of 2D targets and 3D models for the display of digital content such as videos, images, sounds and text related to the content exhibited in the museum, these targets or markers are previously registered in the Vuforia database. for the activation of the user license. On the other hand, with the use of the Unity 3D development engine, the algorithms for the interaction between the user and the superimposed digital content are programmed with the use of the Vuforia SDK, allowing users to have the experience of augmented reality technology. In addition, with the integration of the Flutter SDK, the software integrates an intuitive and friendly graphical user interface, seeking to retain users (see figure 1).

4.2. Software flow diagram

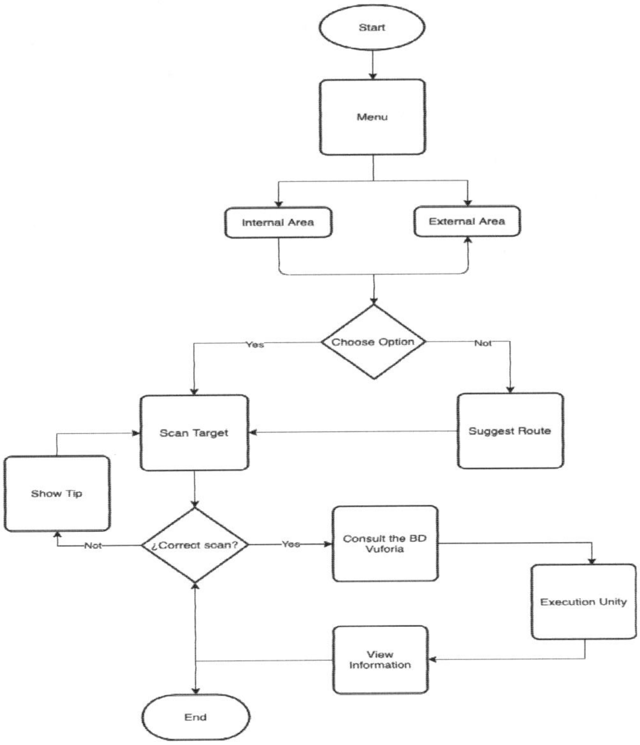

Figure 2. Flow Diagram

In turn, the correct functioning of the mobile application starts from the initial actions of the user and the responses of the application comprised in a series of steps and validations which can be seen in the flow diagram (see figure 2).

5. Discussions

Augmented Reality in the framework of museums means an innovative tool necessary to attract the public to new experiences within them, it is important to study this alternative for the reactivation of exhibition spaces for cultural works. It is a technology that integrates digital and audiovisual content with on-point content in the real world. To achieve the objectives within a project using this technology, different development kits are required depending on the application and the study framework.

In this way, when reviewing the methodology for the implementation and use of this technology in a software development project and all that it entails in the framework of the life cycle of a software (requirements analysis, design, development, implementation and maintenance), it should be noted that it is not necessary to follow this cycle to the letter, this means that integrating augmented reality into a project can be done by people outside the world of software development. this also added to the amount of video tutorials that can be found on video streaming platforms.

6. Conclusions

By way of conclusion, the increase in the incorporation of Augmented Reality in aspects and forms of exhibition of projects is highlighted thanks to the great advances in scientific research, which, in turn, make the use of these AR technologies more accessible and easier to use by people outside of software development. This is how different private and public organizations have chosen to incorporate AR as a solution to a problem or as an extra for their social activities, thus providing an innovative and striking environment for the people who form part of this experience. Likewise, the rise in the use of hardware devices such as smartphones and / or tablets, laptops or virtual reality glasses and the increase in their processing capacity, have had as a consequence that AR is increasingly immersed in daily activities than before, in one way or another, they were more difficult to access by these technologies, however, despite all this, even with all the technology developed in the 21st century, there are still commercial and educational sectors that do not have access to this type of AR technologies. On the other hand, this study contributed to visualizing a design of a system that integrates AR technologies with the mission of contributing to the digitization, exhibition and appropriation of the cultural content exhibited in the different scenarios of the Infantery Museum Marine Park to conclude, the AR in the department of Sucre.

7. Acknowledgement

The authors of this work want to thank the people and institutions that have made this possible. We owe our thanks first and foremost to the Antonio José de Sucre University Corporation and the GINTEING research group.

References

[1] Presidency, "GOV.CO," 23 03 2020. [Online]. Available: https://id.presidencia.gov.co/Paginas/prensa/2020/Gobierno-Nacional-expide-Decreto-457-mediante-el-cual-imparten-instrucciones-para-cumplimiento-Aislaje-Preventiv-200323.aspx. [Accessed 07 2021].

[2] UNESCO, 18 05 2020. [Online]. Available: https://es.unesco.org/news/unesco-y-icom-preocupados-situacion-museos-del-mundo. [Accessed 07 27 2021].

IOP Conf. Series: Materials Science and Engineering 1253 (2022) 012004 doi:10.1088/1757-899X/1253/1/012004

[3] UNESCO, 28 09 2012. [Online]. Available: http://www.unesco.org/new/es/media-services/single-view/news/museums_and_built_heritage_in_culture_and_development/#. [Accessed 2021].

[4] EV Europe, "EVE Museums and Innovation," 08 25 2021. [Online]. Available: https://evemuseografia.com/2021/08/25/_trashed-3_trashed/. [Accessed 2021].

IOP Conf. Series: Materials Science and Engineering 1253 (2022) 012004 doi:10.1088/1757-899X/1253/1/012004

[5] UNESCO, 04 04 2020. [Online]. Available: https://docs.google.com/document/d/1craVAqBASp6U7sCzy5Fpp66d17hHVPcFVPs124CJ42s/edit. [Accessed 2021].

[6] INNOVAE, 2021. [Online]. Available: https://www.innovae.eu/la-realidad-aumentada/. [Accessed 2021].

[7] UNITY, 03 2018. [Online]. Available: https://docs.unity3d.com/es/2018.4/Manual/vuforia-sdk-overview.html. [Accessed 2021].

[8] UNITY, 2021. [Online]. Available: https://unity.com/es/products/unity-platform. [Accessed 2021].

[9] FLUTTER, 2021. [Online]. Available: https://flutter.dev/docs/resources/faq. [Accessed 2021].

[10] GOOGLE, "ARCore," 2021. [Online]. Available: https://developers.google.com/ar/develop. [Accessed 2021].

[11] UNITY, 2018. [Online]. Available: https://docs.unity3d.com/es/2018.4/Manual/vuforia-sdk-overview.html. [Accessed 2021].

[12] VUFORIA, 2021. [Online]. Available: https://library.vuforia.com/features/images/image-targets/the-physical-properties-of-image-based-targets.html. [Accessed 2021].

[13] VUFORIA, 2021. [Online]. Available: https://library.vuforia.com/features/images/image-targets/the-physical-properties-of-image-based-targets.html. [Accessed 2021].

[14] VUFORIA, 2021. [Online]. Available: https://library.vuforia.com/features/objects/model-targets.html. [Accessed 2021].

[15] VUFORIA, 2021. [Online]. Available: https://library.vuforia.com/features/objects/object-reco.html. [Accessed 2021].

[16] MAG Medina and CL Cantero, "Augmented Reality in Museums for the Preseration of the Zenú ethnoidentity," in Interculturality of Ethnicities in Colombia, Sincelejo, Cecar Editorial, 2018, pp. 30 - 32.

[17] SIMCO, "GOV.CO," 2020. [Online]. Available: http://simco.museoscolombianos.gov.co/Home/Museo?personaJuridicaId=664. [Accessed 2021].

[18] DDR Torres and DMLB Gant, "Multimedia guides with augmented reality in 21st century museums: virtuality as an integral part of the exhibition discourse," MODOS revista de historia da arte, vol. 1, no. 1, pp. 1-10, 2017.

[19] RH Sampieri, CF Collado and M. d. PB Lucio, Research methodology. Sixth edition, Mexico City: McGRAW-HILL, 2014.

[20] Minisciences, Typology Of Projects Qualified As Of A Scientific, Technological And Innovation Character Version 5, 2016.

APPENDIX

1. Introduction.
2. Theoretical framework.
   2.1. Augmented Reality (AR).
      2.1.1. Vuforia Engine.
      2.1.2. Unity.
      2.1.3. Flutter.
      2.1.4. ARCore.

   2.2 Follow-up marker-based.
      2.2.1. Image objectives.
      2.2.2. Objective models.
      2.2.3. Objective Objective

   2.3 Museums and Augmented Reality
      2.3.1. Immersive experiences in showrooms

3. Methodology.
4. Results.
   4.1. Software Operation Mapping.
   4.2. Software flow diagram.

5. Discussions.
6. Conclusions.
7. Acknowledgement.

Production of an electrolyte drink from the use of tamarind fruit (*Tamarindus indica L.)*

C.R. Vidal-Tovar [1], Y. Gordon-Hernández[2], P. J. Fragoso-Castilla[3], C.A. Gutierrez De Piñeres[4], and G. E. Angulo-Blanquicett[5]

1 Grupo de Investigación en Sistemas Socio productivos – Universidad Popular del Cesar. Valledupar, Cesar, Colombia.
2. Grupo FACEUPC, Universidad Popular del Cesar. Universidad Popular del Cesar. Valledupar, Cesar, Colombia.
3 Grupo de Investigación en Parasitologia - Agroecologia Milenio Universidad Popular del Cesar. Universidad Popular del Cesar. Valledupar, Cesar, Colombia.
4 Coordinación Extensión Universidad de Santander– UDES, Valledupar, Cesar, Colombia.
5 Grupo de Investigación CIEMPIES, Universidad de Santander, Valledupar, Cesar, Colombia.
carlosvidal@unicesar.edu.co

Abstract. The tamarind (Tamarindus indica L.), is from the legume family and is native to the tropics. The fruit is curved, which the shell is bright brown and its flattened oval seeds, joined together by fibers. It is a highly rustic fruit tree, since it can thrive in poor or marginalized soils, with little or no irrigation and minimal care, in relation to other tropical fruit trees. The objective of this work was to establish the formulation process to obtain a hydrating drink based on Tamarindus indica L., To obtain the drink, the following formulations were made; formulation 1 (6% tamarind pulp - 94% H2O), formulation 2 (9% tamarind pulp - 91% H2O) and formulation 3 (12% tamarind pulp - 88% H2O), keeping the amounts of electrolytes and carbohydrates constant added to the formulation. As the main result for the formulation of the drink, there is the following sequence: Harvesting of the fruit, Selection and classification of the raw material, Pulping, Storage of the pulp, Formulation, Mixing, Pasteurization, Packaging and Storage.

Keywords: Carbohydrates; Electrolytes; Energizing; Hydrating drink; Tropical fruit.

1. Introduction

The tamarind *(Tamarindus indica L.*), is from the legume family and is native to the tropics. The fruit is curved, which the shell is bright brown and its flattened oval seeds, joined together by fibers. It is a highly rustic fruit tree, since it can thrive in poor or marginalized soils, with little or no irrigation and minimal care, in relation to other tropical fruit trees [1, 2]. The fruit is a pod of 12cm to 15cm in length, of which, the pulp constitutes from 30 to 55%, the shell and the fiber from 11 to 30% and the seeds, from brown colour, from 33 to 44%. The harvest is carried out when the fruit reaches its maturity, that is, it becomes dehydrated, loses weight and acquires a brown or brown grey colour and a hollow sound, when the seeds collide with each other. Likewise, the fruit peel becomes brittle when lightly pressed with the fingers [3,4].

Tamarind is present on 4 continents and is cultivated in 54 countries; it is native to 18 nations. Tamarind cultivation is divided into 2 basic categories, sweet and acidic. The acid type constitutes about 95% of the total world production. Even though most tamarind producing countries do not grow on a commercial scale, it has expanded in the last decade [5, 6].

In this way Tirado et al. (2014) [7], reports that fresh consumption of tamarind in the Caribbean region has increased over the years, reaching 1570 tons in 2010. In the same way, the Caribbean region, added to its individual departmental goals. They give 47,550 new hectares to sow in 12 species, which tamarind has 950 hectares, with more land in the department of Cesar with 500 hectares [8].

Therefore, it is observed that in our region there is a wide cultivation of this fruit, but according to Pájaro et al. (2018) [9], it is not being used to make new products, but its most common uses are only for the confectionery industry, the objective of tamarind is the production of pulp for use in the preparation of sweets, juices, jellies, sweetened pulps, bottled soft drinks, ice cream and sauces.

Currently, there are several brands of hydrating drinks designed especially for athletes or for all those who perform regular and intense physical activity [10]. When exercising, much of the energy we spend is converted into heat, if the body did not have the capacity to eliminate it, the body would experience lethal temperatures, for this reason, the body sweats to maintain stable body temperature, and therefore There is so much dehydration, so it is here that hydration gains strength which is key in physical performance [11]. The interest in hydrating beverages has a great impact, since today there is a need to obtain products that do not contain chemical substances that over time can have negative effects on people [12].

Drinking isotonic drinks promotes sports performance and helps to quickly replace losses caused by intense physical exercise. During training, and even during sports competition in a hot and humid environment [13, 14]. But previous research reports that "isotonic" moisturizers are high in artificial substances that cause health effects if they are ingested frequently [15]. For all of the above, the objective of this work was to establish the formulation process to obtain a hydrating drink based on *Tamarindus indica L.*

2. Material and methods

2.1 Research type and design

Quantitative research adopts a systematic, objective, and rigorous strategy to generate and refine knowledge. In this design, deductive reasoning and generalization are used initially. Deductive reasoning is the process in which the researcher begins with an established theory or structure, where concepts have already been reduced to variables, collecting evidence to evaluate or test whether the theory is confirmed. Generalization is the extent to which conclusions developed from the evidence collected in a sample can be extended to a larger population. From the above, this research has a quantitative approach, since experimental results are obtained by the variation of the concentrations of tamarind pulp in different formulations to obtain a hydrating drink, the data in each of the formulations will be statistically analyzed to interpret them and choose the best formulation taking into account the values obtained in the bromatological and physicochemical characteristics.

2.2 Experimental Design

For the three formulations, the variation of pulp and water concentration was taken into account. As constant values the concentration of glucose, sucrose and electrolytes (sodium, calcium, potassium, magnesium and chloride) according to the requirement of the NTC 3837 standard [16], for which 3 liters of each formulation will be prepared for 2 repetitions.

For the concentration of tamarind pulp, it was carried out according to Resolution 3929 of 2013 [17], which requires as a minimum percentage of tamarind fruit in fruit drinks considered nectar the concentration of 10%, this pattern was taken as a reference for vary the values in each formulation of the hydrating drink by 6%, 9% and 12%. Therefore, the variation of the amount of water present in the formulation of each established drink was 94%, 91% and 88% respectively.

On the other hand, NTC 3837 [16] indicates that the maximum percentage of carbohydrates that this class of drink must have is 6% w / v. Therefore, constant values for each formulation of 2% glucose and 2% sucrose were chosen. Likewise, this norm requires some standard parameters for the use of electrolytes, therefore, for the formulation of this drink, percentages in mEq / L were taken for its elaboration: sodium (15), calcium (1.5), potassium (3), chloride (11) and magnesium (0.7), which were kept constant in each of the formulations. These values were taken taking into account that the fruit contributes part of these electrolytes, where they were balanced according to the aforementioned standard, which specifies the limits of each one, therefore approximately average values of each electrolyte were taken. Next, in Table 1, the formulations for the beverage are shown.

Table 1. Formulations for the elaboration of the hydrating drink

Raw material	Formulation 1	Formulation 2	Formulation 3
Pulp (% m/m)	6	9	12
Water (%)	94	91	88
Glucose (%p/v)	2	2	2
Saccharose (%p/v)	2	2	2
Sodium (mEq/L)	15	15	15
Chloride (mEq/L)	11	11	11
Potassium (mEq/L)	3	3	3
Calcium (mEq/L)	1.5	1.5	1.5
Magnesium (mEq/L)	0.7	0.7	0.7

The tamarind fruit was obtained in the supply center of the city of Valledupar. For each formulation, 3 liters were made with 2 repetitions, for which there was a total of 6 liters of the drink. Taking into account the percentages of fruit in each formulation, a quantity of tamarind pulp was required: 360 g (0.36 Kg), 540 g (0.54 Kg) and 720 g (0.72 Kg) respectively. For this, a summation of the total amount of tamarind pulp necessary for the realization of all these formulations was made, which gives an approximate result of 1620 g (1.62 Kg). For this reason, 3 Kg of fruit were purchased, taking into account that in the tamarind there are losses due to waste such as the shells and seeds, which this fruit applied a pulping process to remove the seed and later.

2.3 Organoleptic evaluation of the hydrating drink:

To measure the degree of sensory acceptance of hydrating drink formulations, a sensory analysis was implemented through hedonic testing in 70 untrained tasters belonging to the student population of the Universidad Popular del Cesar, in Valledupar: Each evaluator was given to test 3 samples of drink, one of each formulation with different concentrations of tamarind pulp, at the same time, a survey format was delivered to each evaluator, to record the qualification regarding the product, this person had to observe, smell and taste each of the samples. Then, evaluate the sensory attributes of the product such as taste, smell, texture, color and acceptance, by means of a scale to give a rating ranging from 1 to 4, where 1: "I dislike it", 2: I dislike it little", 3: "I like it" and 4: "I like it very much" [20].

3 Results and Discussion

3.1 Process for preparing the hydrating drink.

Based on the information collected from the theory and background aimed at obtaining hydrating beverages, the technique and methods proposed by Santana (2017) [18] were used, as were the methods proposed by Martelo & Porto (2011) [19] with some adaptations to the techniques and methods outlined above. Harvesting of the tamarind fruit: The tamarind fruit was purchased in the market of the city of

IOP Conf. Series: Materials Science and Engineering 1253 (2022) 012005 doi:10.1088/1757-899X/1253/1/012005

Valledupar, with strict care due to the conditions that the fruit presents to guarantee the quality of the product.

Selection and classification of the raw material: For the selection and classification of the tamarind fruit, the objective was to discard the damaged fruits, to obtain a fruit in optimal conditions, which were used for the process. Once the fruits were selected, the classification was carried out, where they were separated according to their physical appearance, observing that it is free of residues.

Pulping: Once the fruits were selected and classified, the pulping was done to remove the seeds, thus obtaining the tamarind pulp for the preparation of the drink.

Tamarind pulp storage: The pulp was packed in bags where they were stored at a temperature of 7 ° C to preserve the characteristics of the fruit and avoid any type of contamination.

Formulation: Subsequently, the aforementioned formulations were proposed, in which the weighing and preparation of each of the ingredients was carried out.

Mixed: The addition of the components of the drink in established formulations such as water, carbohydrates (glucose and sucrose) and mineral salts (sodium, chloride, potassium, calcium, magnesium), additionally preservatives were added to preserve the product (benzoate of sodium) and acidulant (citric acid).

Pasteurization: A slow pasteurization was carried out at a temperature of 60 ° C for 30 minutes, the rationale for this operation is to reduce the microbial load.

Packaging: A hot packaging was carried out, avoiding the development of microorganisms that could resist heat treatment. The hydrating drink was packaged in glass jars.

Storage: The drink was brought to refrigeration at a temperature of 4° C, to ensure product preservation. Next, Figure 1 shows the flow chart of the process for preparing the hydrating drink.

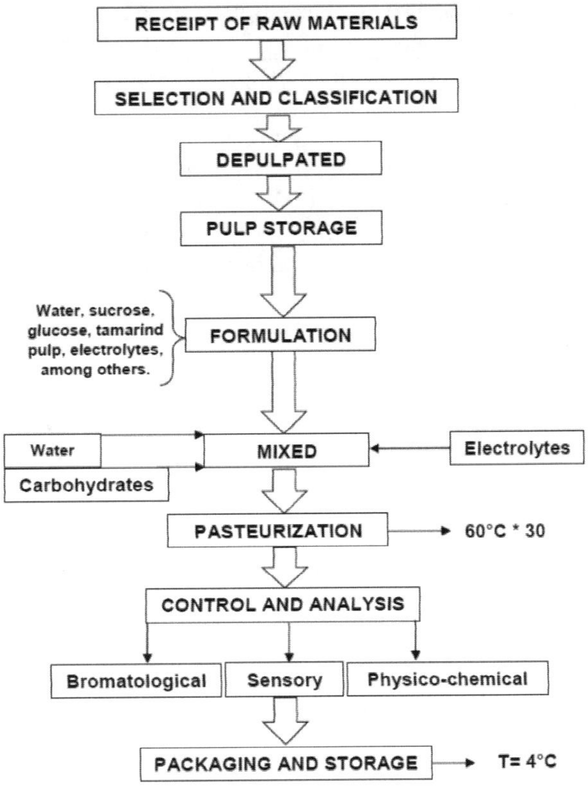

Fig. 1. Flowchart of the process for preparing the hydrating drink.

3.2 Carbohydrate content.

The analysis of variance (ANOVA) table 2 decomposes the variance of glucose and sucrose into two components: an inter-group component and a within-group component, where the ratio-F for glucose, which in this case is equal to 1.08 and the ratio-F for sucrose is equal to 2.63.

Table 2. ANOVA for glucose and sucrose by formulation

Carbohydrates	*Source*	*Sum of Squares*	Df	*Square Mean*	*Ratio-F*	*P- Value*
Glucose	Between groups	149.94	2	74.97	1.08	0.44
	Intra groups	207.67	3	69.22		
Sucrose	Between groups	653.92	2	326,96	2.63	0.21
	Intra groups	373.37	3	124.45		
	Total	1384.90	10	-	-	-

Since the P-value (0.4425) of glucose and the P-value (0.2191) of sucrose is greater than or equal to 0.05, there is no statistically significant difference between the mean glucose and sucrose between one formulation level and another, with a 95.0% confidence level.

Likewise, a multiple comparison procedure is applied to determine which means of glucose are significantly different from others. There are no statistically significant differences. The method used to discriminate the difference between means is Fisher's minimum significant difference (LSD) procedure. It was determined that there were no significant differences for the formulations of 6%, 9% and 12%, with mean values of 19.15, 22.6 and 31.05 respectively. According to Franco et al. (2014), 100 g of tamarind pulp provides approximately 62 g of total carbohydrates, which influences the moment of determining the glucose concentration in the formulations.

Despite the concentration of glucose added (20g), in formulation 1 (6%) the value of glucose remained stable to the amount added with a value of 20.5, while in formulations 2 (9%) and 3 (12%) an increase of this same is noted, so it is evident that the glucose content increases with the increase of tamarind pulp. In addition, Santana (2017), which recorded values from 6.11% to 9.55%, data higher than those obtained in the present study (2.05%, 2.26% and 3.10%), is taken as a reference. This difference can be influenced by the different concentrations and characteristics of the raw materials used in the preparation of the moisturizing drink.

According to the standard NTC 3837 for drinks and energy for physical activity, exercise and sports, provides a range of up to 6 % W/V, expressed as glucose, therefore the glucose values obtained in this investigation are within the acceptable range for a drink moisturizer with values of 2.05%, 2.26% and 3.10%, respectively. On the other hand Martelo and Porto (2011), suggest that a greater amount of carbohydrates translates into an important caloric intake for those who consume these drinks, therefore of all the glucose values obtained in this research, formulation 3 is the most suitable for the preparation of the drink to obtain the highest content (3.105%).

This value obtained in the formulation 3 of glucose (3,105%), reflecting that the carbohydrate content is below other hydrating drinks available on the market, some brands have a percentage of carbohydrates, 5%, other one is with 4%, which reflects that the rehydrating drink made from tamarind can provide the carbohydrates needed as a source of heat for those who consume it.

3.3 Sodium content

The ANOVA table decomposes the variance of sodium into two components: an inter-group component and a within-group component. The ratio-F, which in this case is equal to 123.902, is the ratio between the inter-group estimate and the within-group estimate. Since the P-value of the F-test is less than 0.05, there is a statistically significant difference in the mean sodium between one formulation level and another, with a 95.0% confidence level.

Table 3. ANOVA for sodium by formulation

Source	*Sum of Squares*	Df	*Square Mean*	*Ratio-F*	*P- Value*
Between groups	8.09	2	4.04	123.90	0.0013
Intra groups	0.09	3	0.03		
Total	8.19	5			

Similarly, a multiple comparison procedure is applied to determine which means are significantly different from others. Since the amount of sodium initially added was 15 mEq/L, an increase of sodium was observed in all formulations. For formulation 1 (6%) it recorded a value of 15.01 mEq / L, followed by formulation 2 (9%) which recorded an increase to 16.165 mEq/L and finally formulation 3 (12%) with the highest recorded value of 17.84 mEq/L. According to Franco (2014), the increase is due to the amount of sodium present in 100 grams of tamarind pulp is 28 mg and there is a direct relationship of increase in sodium content by amount of tamarind pulp added to the drink.

In this order, Santana (2017), states that sodium is the only electrolyte that added to sports drinks, to be consumed during physical activity, meets physiological benefits, also mentions that a sodium concentration of 20 to 50 mEq/L, causes the maximum arrival of water and carbohydrates to the small intestine, also allows the maintenance of the volume of extracellular fluid.

From the above, the values of the electrolyte sodium in the drink fluctuate in the range established by the Technical Standard Colombiana NTC 3837 for drinks and energy for physical activity, exercise and sport, which sets a range between 10 – 20 mEq/L. Due to that all the formulations showed significant differences among themselves, it is stated that the best formulation for the preparation of the drink moisturizer is the 3 (12%), as they have the highest amount of sodium (17.84 mEq/L) to allow a greater hydration. Similarly, formulation 3 (17.84 mEq / L) in terms of its value in milligrams (mg) in a portion of 240 ml of drink is 98.47 mg, which is below other moisturizing drinks present on the market with a content between 110 to 125 mg approximately. Paragraph after a heading is not indented (Bodytext style).

3.4 Chloride content

The analysis of variance to determine the effect of chloride in the 3 formulations proposed in the Table 4 it was observed that there is a statistically significant difference in the mean of chloride, between one level of formulation and another, with a level of 95.0% confidence, with a P-value (0.0202) which is lower than the value of the level of significance of the test 0.05.

Table 4: ANOVA for chloride by formulation

Source	*Sum of Squares*	Df	*Square Mean*	*Ratio-F*	*P- Value*
Between groups	2.12	2	1.06	18.74	0.020
Intra groups	0.17	3	0,06		
Total	2.29	5			

According to the multiple ranges test, in which a multiple comparison procedure is applied to determine which chloride means are significantly different from others, it was observed that the interaction (6 – 12) and (9 – 12) presented significant differences with a value of (-1.4 and -1.05). On the other hand, the interaction (6 – 9) did not present a significant difference with a value of (-0.35). You should keep in mind that the tamarind pulp does not provide a considerable amount of chloride, it is possible to show that the chloride in the different formulations presented a decrease in its value-added initially (11 mEq/L), therefore it can be stated in accordance with Guadalupe (2010), its content in the drink depends on the purity of the chloride used in the initial formulation.

Because the interactions (6 – 12) and (9 – 12) presented significant differences, it can be concluded with respect to the multi-range test that formulation 3 with 10.05 mEq/L of chloride is the best for the elaboration of the beverage by showing better results than the other formulations. Similarly, the final content of chlorides of the final drink in the formulation 3 is the only one that coincides with what is established by the Technical Standard Colombiana NTC 3837, which requires the drinks and energy for physical activity, exercise and sport a range of 10 – 12 mEq/L, which complies with the above. Likewise, the chloride content (mg) for formulation 3 (10.05 mEq/L) present in 240 ml of drink is 88.19 mg, a value lower than that presented in some commercial drinks, which are close to 100 mg, but others commercially reach only, however, the values of 36 mg.

3.5 Potassium content.

According to the analysis of variance to determine the effect of potassium on the 3 formulations raised, in Table 5 shows that there is a statistically significant difference between the average of potassium between the level of development and the other, with a level of the 95.0% confidence, taking into account that the P-value (0.0003) is less than the value of the significance level of the test at 0.05, so that at least one of the formulations have significant difference with the other.

Table 5. ANOVA for potassium by formulation

Source	*Sum of Squares*	Df	*Square Mean*	*Ratio-F*	*P- Value*
Between groups	46.29	2	23.14	322.98	0.0003
Intra groups	0.21	3	0.071		
Total	46.50	5			

In that sense, it applies a procedure of multiple comparison to determine which half of potassium are significantly different, it was observed that the formulations of 6%, 9% and 12%, obtained average values of 9,35, 12,55 and 16,15 mEq/L. Given that the amount of potassium added was constant: 3 mEq/L, one can observe an increase in the three formulations due to the potassium content present in 100 g of pulp of tamarind which is of 630 mg. Of this form, the formulation 3 recorded the highest concentration of potassium, in comparison with the other two formulations.

The average values observed in all the formulations differ with the views expressed by the Standard Technique Colombia NTC 3837 for drinks and energy for physical activity, exercise and sports, which states that the maximum value of the potassium in a rehydrating drink should be of 5 mEq/L. however, the results obtained in the investigations of Murillo Calderon (2015), Martelo & Porto (2011), and Guadalupe (2010), recorded values similar to those of this research between 7.797 to 26.49 mEq/L. below the content of moisturizing drinks positioned on the market that have a potassium content between 30 to 43 mg. It should be noted that the potassium content is provided naturally by tamarind, which guarantees reformulate below 3 mEq / L later for other work.

3.6 Organoleptic acceptance

The degree of acceptability of the hydrating drink is analyzed from the information of each of the attributes proposed in Table 6.

The test hedonic acceptance of the rehydrating drink in terms of taste presented statistically significant differences ($p<0.05$) with a confidence level of 95,0% for formulation 1, with respect to the formulations 2 and 3, which showed no significant differences between average, the scores were; 3,26 b ± 0,70 σ and 3,60 b ± 0,50 σ, respectively, by placing the samples within the range of "I like", which reflects that the product has a flavor suitable for consumption .

Table 6. Organoleptic characteristics of the moisturizing drink

Attributes	Drink		
	Formulation 1	Formulation 2	Formulation 3
Flavor	2,46^{a} ± 0,63 σ	3,26^{b} ± 0,70 σ	3,60^{b} ± 0,50 σ
Odour	2,26^{a} ± 0,45 σ	3,60^{c} ± 0,63 σ	2,86^{b} ± 0,83 σ
Texture	2,80^{a} ± 0,56 σ	3,20^{ab} ± 0,67 σ	3,40^{b} ± 0,50 σ
Colour	2,46^{a} ± 0,51 σ	3,06^{b} ± 0,59 σ	3,33^{b} ± 0,61 σ
General acceptance	2,60^{a} ± 0,50 σ	3,20^{b} ± 0,56 σ	3,66^{c} ± 0,48 σ

a-b-c Means with the same letters between columns do not present significant differences. ($p \geq 0.05$).

For the odor attribute, statistically significant differences were observed ($p<0.05$) with a confidence level of 95.0%, in relation to formulations 1, 2 and 3, with formulation 2 having a higher rating of 3.60 c ± 0.63 σ, being located within the "like" scale, demonstrating that the product has a pleasant odor.

In the same sense, for the texture attribute, the observation given by the tasters in this research shows that there is no statistically significant difference between the mean texture ($P \geq 0.05$) between one level of treatments and another, with a 95.0% confidence level. Noting that of the data acquired for this attribute, formulation 2 is related between treatments, for formulations 1 and 3 which are different, with formulation 3 having the highest score with 3.40 b ± 0.50 σ.

On the other hand, the color attribute is presented statistically significant differences ($p<0.05$) with a confidence level of 95.0% for formulation 1, with respect to the formulations 2 and 3, which showed no significant differences between average, the scores were; 3.06 b ± 0.59 σ and 3.33 b ± 0.61 σ, respectively, by placing the samples within the range of "I like", which reflects that the product looks nice visual to the people.

For the general acceptance is presented to a lower grade ($p<0.05$), clarifying that there is significant difference between the average, with a confidence level of 95.0% between formulations 1, 2 and 3 (See figure 1), leading to the formulation 3 the highest rating (3.66 c ± 0.48 σ), followed by the formulation 2 (3.20 b ± 0.56 σ), these values are consistent with the above characteristics, where the formulation 1 presented the lowest score with respect to the overall rating of the attributes flavor, smell, texture, and color.

It is important to point out that the drink is hydrating, formulations 2 and 3 received the most ratings with higher points, indicating that in general these two formulations have good acceptance by consumers, making it clear that the formulation 3 was the best according to the evidence bromatological and physicochemical carried out, confirming so, what is indicated in this test, hedonic, where in general it can be said that the best formulation for rehydrating drink from the fruit of the tamarind, is the formulation 3 with a 12% tamarind pulp.

4. Conclusion

In the present research, a moisturizing drink of natural character was obtained with the appropriate compositions to replenish the electrolytes that are lost during sweating in physical activities. Initially, formulations were proposed at different concentrations (6% tamarind pulp-94% H2O), (9% tamarind pulp – 91% H2O) and (12% tamarind pulp – 88% H2O), then the process was carried out to make each of these formulations according to the provisions of this work. We performed the analysis to the different formulations and chose the language that is closest to the requirements established by the Technical Standard Colombiana NTC 3837 for drinks and energy for physical activity, exercise and sport, being the formulation 3 the one that best meets these requirements, presenting the following compositions: 3,105% glucose, sodium 17,84 mEq/L, chloride 10,05 mEq/L, potassium 16,15 mEq/L and magnesium 7,25 mEq/L, indicating that you got a product with high levels of minerals such as potassium, although these stand out from what is required by the standard, it is taken into account that the largest amount was naturally provided by the tamarind fruit in this research.

References

[1] J. Rodríguez, A. Lafourcade, J. Escalona, "Formulation of tamarind tablets by the wet granulation method", Revista Cubana de Farmacia, vol. 45, no. 3, pp. 414-422, 2011.

[2] J. Viveros, K. Figueroa, F. Gallardo, E. García, O. Rosado, F. Hernández, "Tamarindo (Tamarindus indica L.) management and marketing systems in three municipalities of Veracruz," Revista Mexicana de Ciencias Agrícolas, vol. 3, no. 6, pp. 1217-1230, 2012.

[3] D. Acevedo, D. Tirado, L. Guzman, "Deshidratación Osmótica De Pulpa De Tamarindo (Tamarindus Indica L.): Influencia De La Temperatura Y La Concentración," Revista U.D.C.A Actualidad & Divulgación Científica, vol. 17, no. 1, pp. 123-130, 2014.

[4] N. Islam, A. Liza, O. Faruk, A. Habib, S. Hiziroglu. "Formulation and characterization of tamarind (Tamarindus indica L.) Seed kernel powder (TKP) as green adhesive for lignocellulosic composite industry," International Journal of Biological Macromolecules, vol. 142, pp. 879-888, 2019.

[5] R. Snehashis, M. Chamallamudi, K. Nandakumar, R. Shetty, M. Gill, S. Sumalatha, R. Devkar, K. Gourishetti, N. Kumar, "Evaluation of the aphrodisiac potential of a chemically characterized aqueous extract of Tamarindus indica pulp," Journal of Ethnopharmacology, vol. 210, pp.118-124, 2018.

[6] N. Razali, S. Mat-Junit, A. Faizah, S. Subramaniam, A. Abdul-Aziz, "Effects of various solvents on the extraction of antioxidant phenolics from the leaves, seeds, veins and skins of Tamarindus indica L.," Food Chemistry, vol.131, no. 2, pp. 441-448, 2012.

[7] D. Tirado, D. Acevedo, P. Montero, "Rheological properties of sweetened tamarind pulp," Revista U.D. C. A Actualidad & Divulgación Científica, vol. 17, no. 2, pp. 495-501, 2014.

[8] V. Álvarez, S. Muriel, N. Osorio. "Plants associated with tourism and traditional management systems in the near west Antioquia (Colombia)," Environment and Development, vol. 19, no. 37, pp. 67-82, 2015.

[9] H. Pajaro, J. Benedetti, L. Garcia, "Physicochemical and Microbiological Characterization of a Fruit Wine based on Tamarind (Tamarindus indica L.) and Carambola (Averrhoa carambola L.)," Technological Information, vol. 29, no. 5, pp. 123-130, 2018.A.

[10] Ehlers, G. Marakis, A. Lampen, K. Ildico Hirsch-Ernst, "Risk assessment of energy drinks with focus on cardiovascular parameters and energy drink consumption in Europe," Food and Chemical Toxicology, vol. 130, pp. 109-121, 2019.

[11] G. Llanos, J. Piedrahíta, S. Sierra, C. Flórez, L. Acosta, "Elaboration of an energy drink from borojó (Borojo apatinoi Cuatrec.), "Journal Lasallian Research, vol. 9, no. 2, 33-43, 2012.

[12] A. Cuellas, J. Wagner, "Preparation of energy drink from cheese whey," Innotec, vol. 5, no. 1, pp. 54-57, 2010.

[13] M. González, S. Mauro, B. García, D. Fajardo, E. Garicano-Vilar, (2015). "Nutritional assessment, evaluation of body composition and its relationship with athletic performance in a women's soccer team," Revista Española de Nutrición Humana y Dietética, vol. 19, no. 1, pp. 36-48, 2015.

[14] A. Estévez, "Nutritional planning for an amateur swimmer", Journal of Education, Motor Skills and Research, vol. 3, pp. 34-69, 2014.

[15] A. Urdampilleta, J. Martínez, S. Sanchez, J. Álvarez, "Hydration protocol before, during and after physical-sports activity. Motricity, " European Journal of Human Movement, vol. 31, pp. 57-76, 2013.

[16] NTC 3837 "Colombian Technical Standard, Non-alcoholic Beverages. Hydrating Drinks for Physical Activity and Sport, (Online). Retrieved August 8, 2019.

[17] Resolution 3929, "Technical regulation on the sanitary requirements to be met by fruits and beverages with the addition of juice (juice) or fruit pulp, 1-29, Bogotá, Colombia (2013)

[18] P. Santana, "cocoa Mucilage (Theobroma Cacao L.), National And Trinitarian For Obtaining A

Moisturizing Drink,". State Technical University of Quevedo, 2017.
[19] M. Martelo, T. Porto, "elaboration Of A Moisturizing Drink Based On Carambola (Averrhoa Carambola L.)," University of Cartagena, 2011.
[20] C.R. Vidal Tovar, C.A. Cadavid Osorio, G.E. Angulo Blanquicet and C.A. Severiche-Sierra. Obtaining a sugar syrup from the use of the extract of the guácimo fruit (Guazuma ulmifolia lam). IOP Conference Series: Materials Science and Engineering. 2020. Tomado de: https://doi.org/10.1088/1757-899X/844/1/012003

IOP Conf. Series: Materials Science and Engineering 1253 (2022) 012006 doi:10.1088/1757-899X/1253/1/012006

Design of a software prototype to make a Tour in the Infantery Museum Marine Park using View 360 technology.

E. Guzman[1], C. Hernandez[1], A. Morales[1], S. Sánchez[1] and M. Acurero[2]

[1]Antonio José de Sucre University Corporation (Colombia).

[2] Caribbean University Corporation – CECAR (Colombia)

docente_investigador6@corposucre.edu.co

Abstract. Before the arrival of the pandemic due to SARS-CoV-2, the Infantery Museum Marine Park was among the most visited tourist sites in Coveñas, however, due to Colombian decree 457, the museum closed its doors to the visitors, thus avoiding the exposure of the national military heritages that are exposed in it, restricting one of the most important contributions to the educational context in the Caribbean Region. Therefore, the objective of the research is to design the model of a web prototype, using View 360 ° technology, for the remote exhibition of the facilities and heritage of the Infantery Museum Marine Park. The type of research is basic and applied, where a documentary analysis of the 360 ° view technology was carried out, identifying the technical and user requirements for the development of the prototype design, which will integrate 360 ° panoramic photographs and points of access, strengthening the exhibition of the heritages and there was a completely new way for the museum professionals to communicate the exhibits to the public. It is concluded that the proposed prototype meets the functional requirements necessary to solve the problem presented.

1. Introduction

The pandemic generated by COVID 19 brought as a consequence, the closure of different public and private establishments, since worldwide communities have been asked to stay at home and avoid large crowds of people to Soprevent the spread of the virus. An example of these establishments are museums, whose purpose is to communicate, conserve, acquire, investigate, expose and exhibit, the educational material of the historical cultural heritage. It is considered that 90% of the approximately 60,000 museums in the world have been forced to close their doors partially or totally to the general public due to the epidemic. For this reason, museums have chosento implement new technologies to address this challenge and continue to provide information on world heritage to generate new knowledge [1].

The new technologies that are currently available are diverse, as well as their applications in various fields, such as multimedia that is used for different purposes such as tourism, pcultural heritage, archeology, anddducation, the mdigital marketing, among other fields. Its interactivity opens the possibility that multimedia content can be more dynamic and interesting for the user, since it is possible to interact and at the same time keep track of what is being done. In this sense, it is important to determine that there is a number of technologies that provide an immersive sensation such as virtual reality, this type of technology is a simulation of a three-dimensional environment generated by computers, in which the user is capable of both seeing As well as manipulating the multimedia contents of that environment, they generally give the user the feeling as if they were actually inside thecontext real, recording what happens in the scene [2].

The main objective of a VR system is to achieve mental immersion through sensory immersion, for this different types of devices are used, among which VR glasses, sensory gloves, headphones, motion capture suits, stand out. mobile devices, among others. In order to perform a two-way interaction with the user[3].

Virtual reality is the new reality, and a Virtual Tour is a good way to represent it. The Virtual tours or tours are an attractive way of promoting objects or places to the public in 360 ° panoramic photographs (360 ° View Technology), integrated with interactive images, audios, videos, making these virtual visits quite interesting and accessible, since they are can see through the Web and the user have constant control throughout the virtual tour. On an international levelThis type of system is widely used because it generates communication opportunities with greater dynamism, especially it is used by hotels, museums, universities, among other institutions; andBecause, through these virtual tours, the facilities can be presented in such a way as to create visual recognition for the user. To levelnational in Colombia, virtual reality It is being introduced gradually to improve the user experience, thanks to new technologies, it has become more accessible and its use is growing. A virtual museum tour can represent an accessible and participatory space to generate new knowledge, and when applied to cultural heritage, a virtual reality exhibition can provide museum professionals with a completely new and improved way of communicating exhibits. to the public.

Regarding the Infantery Museum Marine Park, which opened its doors to the public on August 17, 2012, created with the aim of spreading the history of the Colombian Marine Corps. Before the arrival of the pandemic due to Covid-19, it had become one of the most famous tourist sites in Coveñas, thanks to the historical legacy that this place has. Providing its visitors with information on the Colombian marine infantry from the time of independence to the current date, using its cultural, educational, technological, recreational and ecological spaces, to expose the evolution of this military force. One of his most important contributions has been the constant visits of the different educational institutions of basic levelsto, mediay superior in the Caribbean Region, since for the Infantery Museum Marine Park it is important to be able sensitize children, adolescents, youth and adults about the military assets of Colombia, so that they know and appropriate the military history, specifically of the country's navy and the Caribbean Region, however many of the educational institutions do not have a financial subsidy necessary to make the displacement of their students to the municipality of Coveñas where the museo, and added to that, the pandemic has made it even more difficult to provide these spaces to different educational establishments to contribute to the training of their students. [4]

Therefore, the main objective of this research work is to design a software prototype to carry out a virtual tour that implements View 360º technology., allowing the visualization of the main objects of the Infantery Museum Marine Park, this to give the general public the possibility of knowing its facilities remotely.

2. Theoretical framework

2.1 Virtual reality

According to Josu Otegui[5], defines virtual reality as a computer system used to create an artificial world, this being generated by a computer, and a device such as a virtual camera allowing different users to view environments similar to real ones, thus allowing manipulation and make an interaction in real time, achieving this through a device which allows to achieve the virtual presence of these environments.

Likewise, Noelia Moreno, Eloy López, Juan Leiva[6], establish that virtual reality as a computer system that produces a computerized simulation and representation of reality, is characterized by its immersive technology, allowing the user to immerse themselves in three-dimensional scenarios in the first person and in 360 degrees.

On the other hand, José Herranz De La casa, Pavel Sidorenko [7], affirm that the concept of virtual reality has undergone some changes over time, but they define this technology as a digital experience enhanced through a vision gadget (VR Glasses), through which both emotional and emotional sensations and reactions are generated. Physics, in the same way as they are experienced in real life, is the illusion of something that happens and exists in reality.

2.2 360 photography

360 degree photography is a method used to capture photos in which the observer can immerse himself in the image so that he can feel inside it. It allows you to see what is happening around the scene, that is, to see in all directions. The environment can be observed as if we were inside it thanks to the high definition and the quality of the images[8].

360 ° photography is a component that is within virtual reality since it allows us to observe the spaces with a total perception and these spaces are an identical recreation of reality. With the novelty of recreating real scenarios of the work environment of a future professional, it is hoped that learning situations are created or designed so that the student uses their knowledge to a possible real situation[9].

2.3 Cultural heritage

On the other hand, Alejandra Cabrera, Antonio Vidal [10], affirm that cultural heritage is a very important tool to generate social cohesion in separate or disintegrated societies, in the same way to promote new social, productive and political relationships, it can also create dynamics of equity, affinity and new forms of appropriation of the territory, that promote participation and the exercise of citizenship.

According to Alicia Fernández, Rocío González [eleven], cultural heritage is made up of physical components such as museums and historical monuments and intangible components such as traditions or language. Museums expose the history of a city or civilization, as well as some cultural heritage elements of a certain period or historical event or similarly focused on a building of avant-garde architecture and art, thus fulfilling an important role in the preservation of this type of heritage.

According to Cazaux, DE (2019) Third generation museums correspond to the majority of interactive science centers. These centers are collections of ideas, natural phenomena and scientific principles that they try to transmit, more than objects. The emphasis of these centers is the active participation of the visitor and their approach is mostly interactive, that is, they seek independence and reciprocal action between the exhibition and the user. They tend to be based on modern technologies and playful approaches. They give primacy to the four-dimensional experience, considering that in exhibitions the objects are three-dimensional and interactivity is the fourth dimension.[12].

3. Methodology

The present study is framed in two types of research, the first type is a basic study of descriptive scope, which according to MinCScience consists of experimental or theoretical works that are undertaken mainly to obtain new knowledge about the foundations, phenomena and observable facts, without having as objective any application or use determined, regardless of the area of knowledge [12]. In this stage the purpose of this research is to study View 360 ° technologies to define a model that serves as a basis or foundation in the design of the prototype. Then the investigation pass to become a type of applied scientific research, which, according to MinCscience. Its main objective is to acquire new technical or scientific knowledge, oriented towards a specific practical objective. To achieve this, all existing and available knowledge to solve specific problems must be considered. In other words, the research applied to this study will allow decidethe use360 ° view technology for design a software modeling whose purpose will be to make a virtual visit to the Infantery Museum Marine Park, for those who want to know the military heritage of the Colombian National Navy.

3.1 Procedures

To carry out the research, two phases were developed using the Cascade methodology, applying the requirements phase and the design phase. (see figure 1):

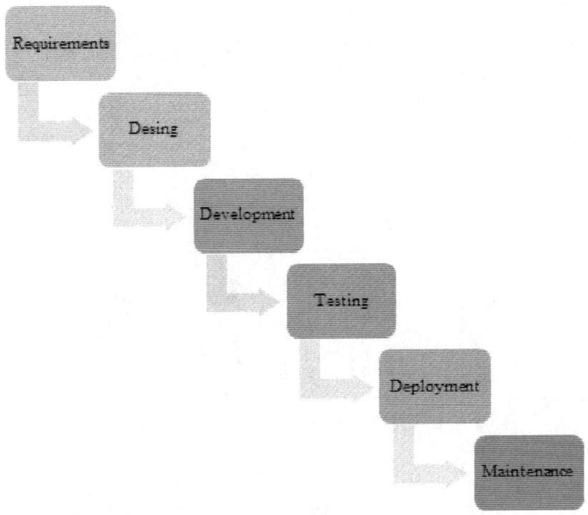

Figure 1. Cascade Methodology.

4. Results

As a result of this research process, a 360 ° virtual reality environment of the Infantery Museum Marine Park was designed, the modeling of the web platform was created using UML diagram, which provide a schematic of the architecture, in that way The design of a virtual tour was created, providing features such as control of the route's handling, information points and access to other points along the route. A bibliographic review was organized in a bibliographic matrix and analyzed by means of an analytical content matrix, later the requirements were raised through semi-structured interviews applied to the administrative staff of the Infantery Museum Marine Park, said requirements are listed in Table 1.

Table 1. Functional and non-functional requirements

Functional	Not functional
The system will allow the user to navigate through the facilities of the Infantery Museum Marine Park.	System interfaces should be easy or intuitive.
The system will allow the user to obtain information about each of the pieces exhibited during the experience, these being: videos, images, audio and texts.	The command system will be with the gyroscope of a smartphone or with the mouse of a computer
The system will allow the user to view and interact with a navigation map during the tour.	The design of the pieces exhibited during the experience must have a good edition and quality images.

IOP Conf. Series: Materials Science and Engineering 1253 (2022) 012006 doi:10.1088/1757-899X/1253/1/012006

Within the virtual tour there is no time limit, so the user can stay unlimited.	The protocol used to handle the security of the system must be reliable enough.
To use the virtual tour, you should not do any type of registration or login.	The application must be hosted on a hosting, and thus be able to be used in the different types of web browsers.

Following the analysis of the requirements, the behavior of the proposed web prototype was modeled, which are displayed below:

Figure 2 shows the case diagram that includes the milestones of the prototype development, subdivided in the design phase.

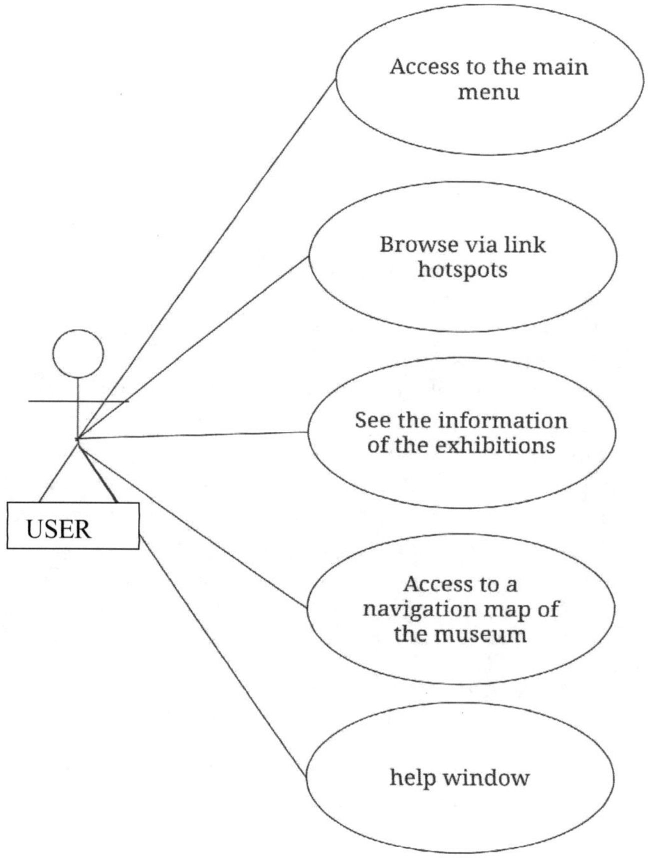

Figure 2. Prototype use case diagram

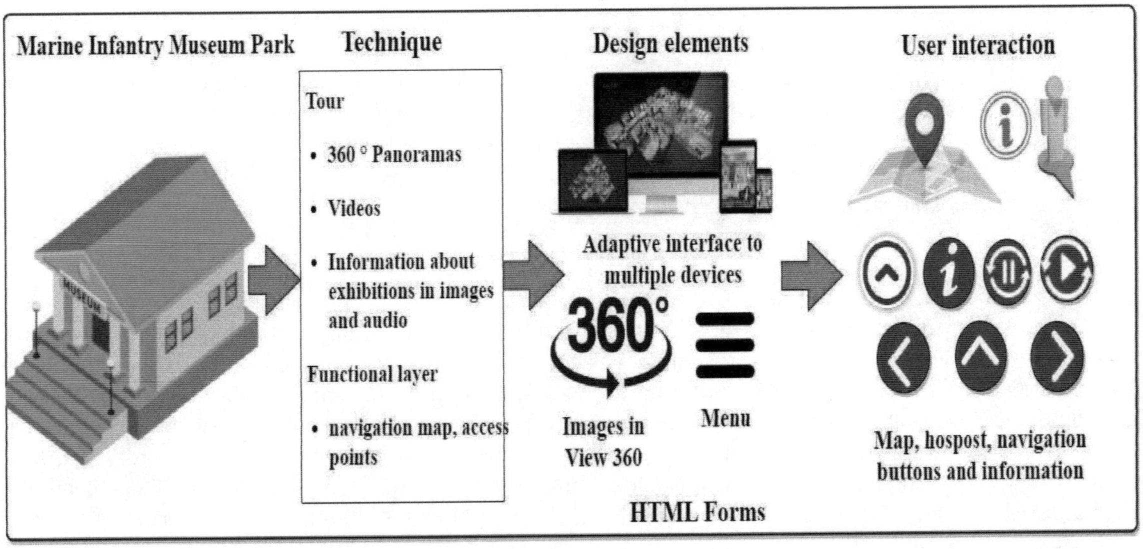

Figure 3. Conceptual model of the prototype

Figure 3 shows the different techniques used to present the virtual tour, as well as information on the exhibits of the Infantery Museum Marine Park. The visual display features for the user include a navigation map, hotspot. Meanwhile, the functional layer of the prototype is used to show 360 ° images and the information in texts and audios of each of the elements of the museum. A design of an interface adaptable to the different devices and a menu was included which will be the access door for the content of the web prototype.

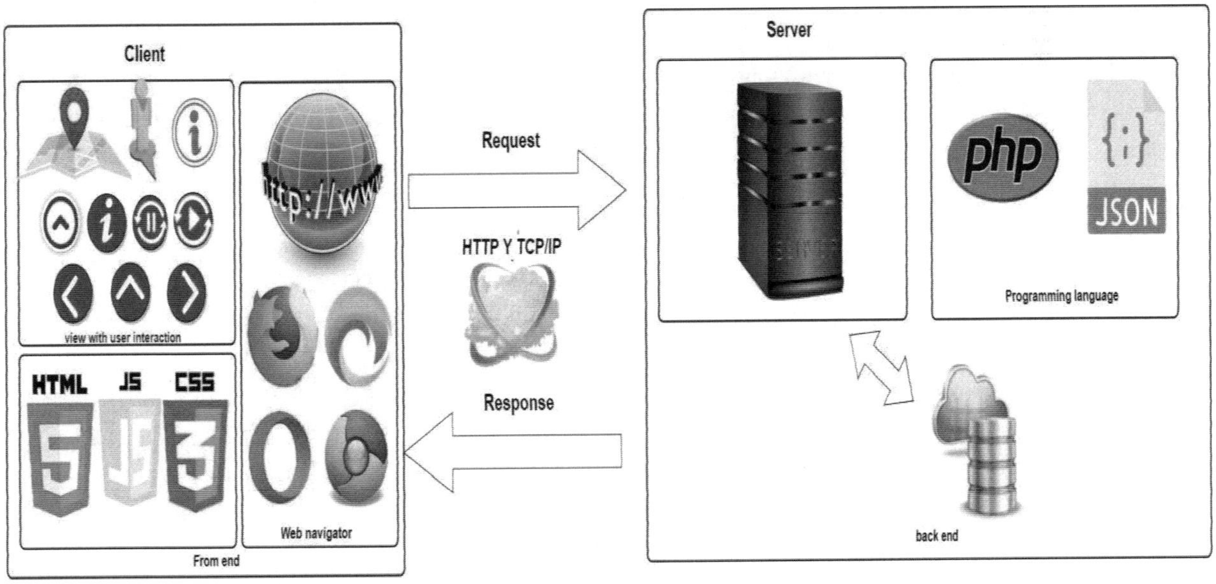

Figure 4. Prototype architecture scheme

This figure 4 represents the client / server architecture in the prototype design, the client is generally represented by a web browser through which it makes requests to the server using the different options of the application's navigation menu, which They include a navigation map, hotspot or access link to other parts of the exhibition and different buttons to listen to informative audios or view illustrative images of texts from each of the exhibitions. On the part of the server, the web application will be represented in which the user's requests will be processed, in response to those requests, different results will be displayed in the browser, such as 360 ° images and information in texts and audios of each of the elements of the museum

5. Conclusions

From the results obtained, it is concluded that the development of applications with virtual reality allows the simulation of events or phenomena that are exhibited within the different scenarios that are shown in the museum, thus facilitating remote interaction by students and private visitors. The main objective of this study being to propose an effective interaction design to develop a web application in order to make a virtual tour of the Infantery Museum Marine Park. In addition, an interview was conducted with the museum's administrative staff to obtain the user's perspective regarding the information display of each exhibited object, thus facilitating a guide booklet of all the museum's exhibits.

References

[1] UNESCO, «UNESCO,» 04 04 2020. [Online]. Available: https://es.unesco.org/news/ museums-challenges-covid-19-remain-committed-communities.

[2] ER Escartín, «VIRTUAL REALITY, AN EDUCATIONAL TECHNOLOGY WITHIN OUR REACH,» [Online]. Available: https://core.ac.uk/download/pdf/51408046.pdf.

[3] MA Gutierrez, F. Vexo and D. Thalmann, "Stepping into Virtual Reality," 2008. [Online]. Available: https://books.google.com.py/books?id=yI8o7osCuQoC&printsec=frontcover&hl=es&source=gbs_ge_summary_r&cad=0#v=onepage&q&f=false.

[4] E. Villalobos, "https://blog.redbus.co," [Online]. Available: https://blog.redbus.co/ tourist-places/park-museum-marine-infantry-covenas /.

[5] J. OTEGUI CASTILLO, «VIRTUAL REALITY AND AUGMENTED REALITY IN THE MARKETING PROCESS,» 12 2017. [Online]. Available: https://addi.ehu.es/bitstream/handle/10810/24910/9.J.Otegui.pdf?sequence=1&isAllowed=y.

[6] NM Moreno Martínez, E. López Meneses and JJ Leiva Olivencia, «The use of emerging technologies as didactic resources in educational settings,» 15 10 2017. [Online]. Available: http://www.hottopos.com/isle29_30/131-146Moreno.pdf.

[7] JM Herranz de la Casa and P. Sidorenko Bautista, «Virtual reality and 360º video in business and institutional communication,» 08 2019. [Online]. Available: https://www.researchgate.net/publication/335585670_Virtual_reality_and_360_video_in_business_and_in stitutional_communication.

[8] DS Barba Ayala, «VIRTUAL CATALOG FOR THE WEB, USING 360 PHOTOGRAPHS OF THE MOST VISITED ATTRACTIONS IN THE CITY OF PUYO,» 2019. [Online]. Available: http://dspace.espoch.edu.ec/bitstream/123456789/11692/1/88T00320.pdf.

[9] LM Trujillo Flórez, «Implementation of the area for immersive environments in 360 photography,» 2017. [Online]. Available: https://journal.poligran.edu.co/index.php/libros/article/view/1941/1867.

[10] AM Cabrera Martínez and A. Vidal Ortega, «Organization of cultural heritage in Colombia: An unexplored category,» 01/27/2017. [Online]. Available: https://poseidon01.ssrn.com/delivery.php?ID=833004024029081104103069119114126030102056089014095061075097086117074090111086000028028097023029026127033124073024013029028032073024013023029036012073024013023022036012073024013023029038024073024013023029036013073024013023029019031240730240130230920 3601

[eleven] A. Fernández Cortés and R. González Sánchez, «Analysis of the use of technology in museums: intelligent museums,» 2019. [Online]. Available: https://www.uco.es/ucopress/ojs/index.php/riturem/article/view/11190/10899.

[12] D. Cazaux, "Origin and development of interactive science and technology museums," 2016. [Online]. Available: https://repositorio.itm.edu.co/bitstream/handle/20.500.12622/1886/Origin%20y%20 developing %20de%20los%20Museum.pdf?sequence=1&isAllowed=y.

[13] «MINCIENCIAS,» 2016. [Online]. Available: https://minciencias.gov.co/sites/default/files/upload/ announcement /anex_1- project_typology_document_version _6.pdf.

IOP Conf. Series: Materials Science and Engineering 1253 (2022) 012007 doi:10.1088/1757-899X/1253/1/012007

Machine learning for population growth modeling through annual CO_2 data

Gustavo Suárez[1], José Daniel Hoyos[2], Esteban Vallejo[3], Julián Andrés Arias[3]

[1]Grupo de Investigación e Innovación en Energía, Institución Universitaria Pascual Bravo, Medellín, Colombia
[2]School of Aeronautics and Astronautics, Purdue University, West Lafayette, United States
[3]Grupo de Investigación sobre Nuevos Materiales, Universidad Pontificia Bolivariana, Medellín, Colombia

E-mail: gustavo.suarez@pascualbravo.edu.co, jhoyos@purdue.edu, esteban.vallejomo@upb.edu.co, julian.ariasf@upb.edu.co

Abstract. This article presents a general framework to model the population growth of Colombia through machine learning, which allows incorporating variables such as CO_2 emissions. The data are obtained through the DANE Geoportal and the World Bank portal database. In recent years, several cities in Colombia have experienced an accelerated population and urban growth as a result of an overwhelming increase in the immigrant population, a situation that has been affecting the urban order of the city, mainly due to population confinement, increased environmental pollution, unhealthiness, poverty among other factors.

The population growth prediction fully shows the effects of the fertility rate, mortality, and migration of people from one territory to another, essential factors in determining population and health system conditions. Accurate information can be obtained from population census measurements, a strategy with a high cost; However, the capabilities of machine learning are growing, being also implemented to determine predictions of systems and behaviors in demographic, environmental, and socioeconomic issues, which contributes to the economic and energy-saving that implies the gathering and processing of demographic information for the country.

In this work, to predict population growth, the concept of machine learning was applied, carrying out a data analysis using an Artificial Neural Network (ANN). The results obtained provide a predictive model with the potential to provide regional and governmental entities with suggestions, strategies, and alternatives for studying the population and making decisions for urban redevelopment.

1. Introduction

The rapid urbanization of cities is a global trend that poses both challenges and opportunities for long-term sustainability. As the population grows and people continue to migrate to urban centers, issues related to poverty, informal settlements, insecurity, air pollution, noise, visual pollution, rapidly spreading diseases, and the economic cost of expanding public infrastructure becomes essential. Cities consume raw materials in large quantities, accelerating environmental degradation and affecting climate change [1, 2].

Population projections estimate the birth rate, death rate, and population migration from one place to another that are used for future predictions. Prediction and forecasting is a complex task where the information of a social system can become chaotic, becoming unpredictable. Furthermore, in most cases, these systems are influenced by climate, pollution, educational, political, and economic conditions of the territory. To predict what cities will look like in the future, urban planners rely on historical data and urban growth models. These models can have multiple outcomes but commonly provide a distribution of the population within a study region over a long period [3, 4].

Employing machine learning the forecasting can be done efficiently. Predictions can be made in two different ways: supervised learning and unsupervised learning. The algorithm is trained on a predefined set of training examples that allow the algorithm to obtain a prediction from a set of data [5, 6].

2. Modelling

A predictive model for the population growth of Colombia was built based on a 61-year history through an ANN Artificial Neural Network with supervised learning.

The dataset was compiled using data available on the Dane [7] and World Bank [8] websites. Using the filter tools available by the World Bank, it was possible to build a dataset that included demographic variables and CO_2 emission indicators in Colombia during the years 1960 and 2020.

2.1. Input variables

To generate the prediction, it was considered to take the annual statistics of CO_2 emissions as a value that allows predicting the number of people who consume and produce goods and services in a given territory [9, 10]. Table 1 describes the variables selected to train the model, with particular emphasis on those variables that concern CO_2 emission indicators.

The dependent variable corresponds to "Population, total", the value to be predicted, while the independent variables correspond to the others. Therefore we have a total of 9 input variables and one output variable. Each record in the database corresponds to a year between 1960 and 2020, which results in a sample of 60 records to carry out the study.

2.2. Libraries

For the implementation code, some libraries were required to facilitate the manipulation, visualization, and treatment of the data, which are presented in Table 2.

2.3. Neuronal network structure

The architecture developed for the ANN consists of an input layer of 9 neurons (corresponding to the nine independent variables), a hidden layer of 4 neurons that connect the input layer with the output layer, which contains a single neuron that represents the selected output variable to predict the total population, Fig. 1.

In the first and second layers, a ReLU activation function was configured, while a linear activation function was established for the output layer. Finally, 80% of the data was used to train and the remaining 20% to evaluate the predictive model.

2.4. Data pre-processing

The input data is manipulated before its process within the neuronal network. The input data is handled before entering the neural network since some data can create issues with convergence or problems of non-derivability, so data that does not meet the conditions should be eliminated from the dataset. Then the data must go through the normalization function as shown in Eq. 1.

Table 1: Demographic and CO_2 pollution variables.

Variable	Description
CO2 emissions from gaseous fuel consumption (kt)	Carbon dioxide emissions from liquid fuel consumption refer mainly to emissions from use of natural gas as an energy source.
CO2 emissions from solid fuel consumption (kt)	Carbon dioxide emissions from solid fuel consumption refer mainly to emissions from use of coal as an energy source.
CO2 emissions from transport (% of total fuel combustion)	CO2 emissions from transport contains emissions from the combustion of fuel for all transport activity, regardless of the sector, except for international marine bunkers and international aviation.
CO2 emissions from electricity and heat production, total (% of total fuel combustion)	CO2 emissions from electricity and heat production are the sum of three CO2 emission categories: (1) Electricity and heat from the main activity producer, (2) Unassigned autoproducers that contain emissions from the generation of electricity and / or heat by autoproducers. (3) Other energy industries that contain emissions from fuels burned in oil refineries, for the manufacture of solid fuels, coal extraction, oil and gas extraction and other energy-producing industries.
CO2 emissions from solid fuel consumption (% of total)	Carbon dioxide emissions from solid fuel consumption refer mainly to emissions from use of coal as an energy source.
CO2 emissions from other sectors, excluding residential buildings and commercial and public services (% of total fuel combustion)	CO2 emissions from other sectors, less residential buildings and commercial and public services, contains the emissions from commercial/institutional activities, residential, agriculture/forestry, fishing and other emissions not specified elsewhere that are included in the IPCC Source/Sink Categories 1 A 4 and 1 A 5. In the 1996 IPCC Guidelines, the category also includes emissions from autoproducers in the commercial/residential/agricultural sectors that generate electricity and/or heat. The IEA data are not collected in a way that allows the energy consumption to be split by specific end-use and therefore, autoproducers are shown as a separate item (Unallocated Autoproducers).
Birth rate, crude (per 1,000 people)	Crude birth rate indicates the number of live births occurring during the year, per 1,000 population estimated at midyear. Subtracting the crude death rate from the crude birth rate provides the rate of natural increase, which is equal to the rate of population change in the absence of migration.
Death rate, crude (per 1,000 people)	Crude death rate indicates the number of deaths occurring during the year, per 1,000 population estimated at midyear. Subtracting the crude death rate from the crude birth rate provides the rate of natural increase, which is equal to the rate of population change in the absence of migration.
Net migration	Net migration is the net total of migrants during the period, that is, the total number of immigrants less the annual number of emigrants, including both citizens and noncitizens.
Population, total	Total population counts all residents regardless of legal status or citizenship. The values shown are midyear estimates.

Table 2: Libraries employed.

Library name	Abbreviation	Use
Pandas	pandas	Data manipulation and analysis
Numeric Python	numpy	Matrix manipulation and numerical analysis
Scikit-learn	sklearn	Tools for programming neural networks
Keras Tensorflow	keras	Tool for simple management of deep learning neural networks
Matplotlib	matplotlib	Graph generation

$$X_{norm} = \frac{X - X_{min}}{X_{max} - X_{min}} \tag{1}$$

Where X is the value to be transformed, X_{min} is the minimum value in the data vector, and X_{max} is the maximum value of the data vector. Thus, this process allows obtaining values between 0 and 1 which is vital for the correct operation of the activation functions.

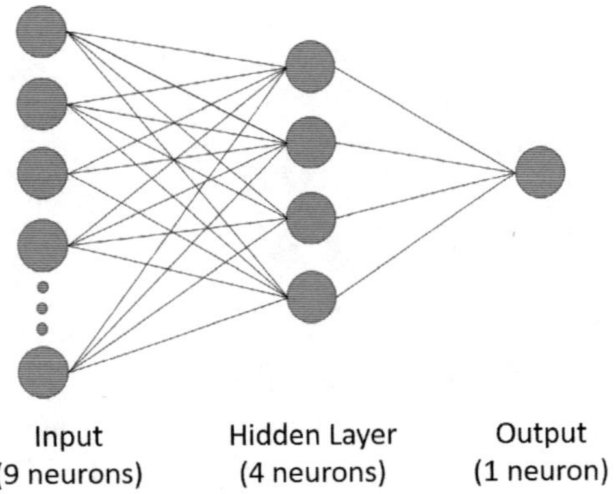

Figure 1: Neuronal network structure.

2.5. Data post-processing
The output data must be re-scaled by solving for X from Eq. 1. Finally, the training data is compared against actual data with a calculation of mean absolute error.

3. Results and discussion

The ANN was successfully trained with the collected demographic and CO_2 emission data. A mean absolute error of 0.038 was obtained in evaluating the model with 20% of the dataset. Fig. 2 establishes a relationship between some input variables of the model on the X-axis and the output variable on the Y-axis. The blue line indicates the behavior of the relationship between the values of the variable on the X-axis and the known values of the output variable (Population, total). The red line indicates the behavior of the relationship between the values of the X-axis variable and the values predicted by the ANN of the output variable (Population, total). The superposition of these behaviors allows analyzing the precision with which the model fits the patterns learned from the training data.

The mean absolute error obtained after the evaluation of the model indicates a favorable capacity of the trained ANN to adjust to the behavior of the collected data. It implies a correct configuration of the network architecture and a satisfactory pre-processing of the data.

Fig. 2 shows a great fit to the behavior of the data selected for the evaluation of the model since the superposition of lines indicates a similar pattern on the part of the values predicted by the trained model. The absence of a perfect overlap generates confidence that the model is not overtrained, which will allow it to receive new data and predict the amount of the Colombian population that logically corresponds to the stimuli of the demographic and environmental variables considered.

The significant adjustment of the variables in figure 2 associated with CO_2 emission and net migration could indicate the relevance of these indicators for the prediction of the population in Colombia, an expected behavior, since health problems are associated with pollution, affecting mortality rates, which decrease the population of a country. The relationship with the exit variable is evident regarding migration since the entry or exit of people to the country increases or decreases its population. It is essential to consider that fluctuations in the number of people in the country affect the emission of CO_2 into the atmosphere, configuring a cycle of relationships between the variables requiring prompt human intervention to break a detrimental dynamic the planet, especially for its inhabitants.

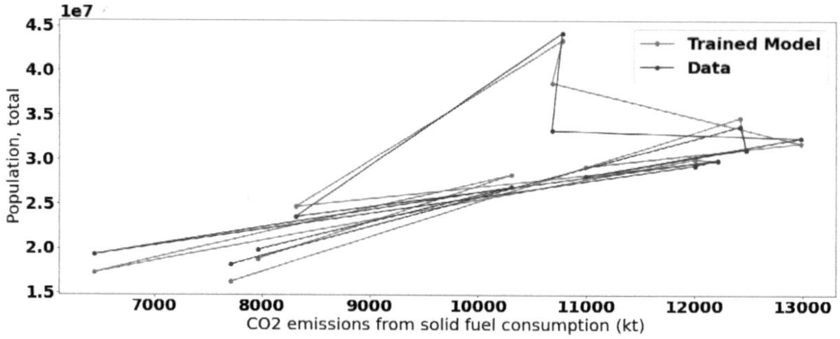

(a)

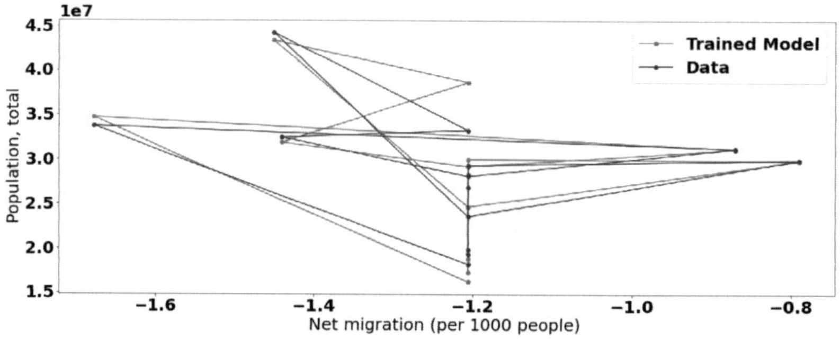

(b)

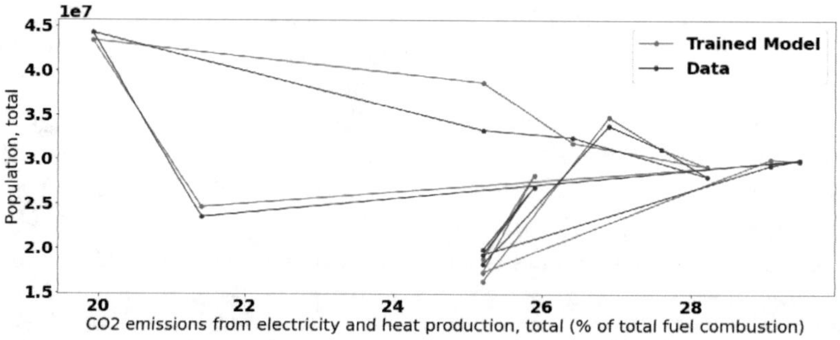

(c)

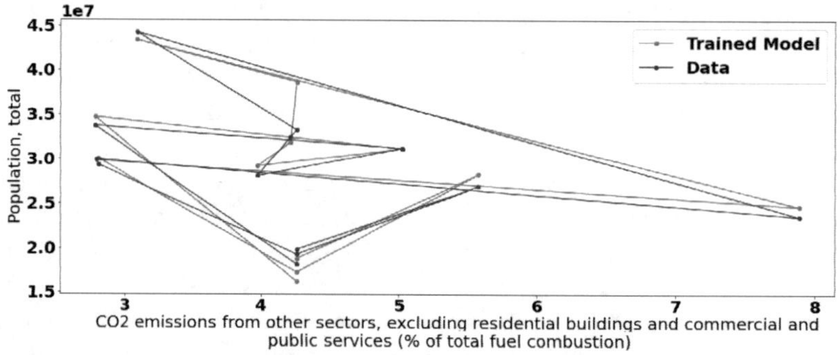

(d)

Figure 2: ANN relationship between Population, total and a) CO_2 emissions from solid fuel consumption (kt), b) Net migration, c) CO_2 emissions from electricity and heat production, total (% of total fuel combustion), d) CO_2 emissions from other sectors, excluding residential buildings and commercial and public services (% of total fuel combustion).

IOP Conf. Series: Materials Science and Engineering 1253 (2022) 012007 doi:10.1088/1757-899X/1253/1/012007

4. Conclusions

Predictive models allow anticipating social, environmental, and health behaviors that can affect the economies of the countries, the urban order, safety, among other demographic factors.

Machine learning tools could contribute globally to strategies, norms, and laws that lead to society's better environmental, economic, and health decisions.

One of the world situations, which in the same way has been affecting Colombia and its regions (Soacha, Bogotá, among others), is the migratory flow which comes mainly from nearby countries and which has been affecting not only the economy but also health and environment. This situation requires governments to implement strategies that provide accurate measurements to guide procedures and plans that mitigate environmental and human health impacts.

References

[1] Gómez J A, Patiño J E, Duque J C and Passos S 2020 *Remote Sensing* **12** ISSN 2072-4292 URL https://www.mdpi.com/2072-4292/12/1/109

[2] Hui J 2020 *bioRxiv* (*Preprint* https://www.biorxiv.org/content/early/2020/07/14/2020.07.14.202119.full.pdf) URL https://www.biorxiv.org/content/early/2020/07/14/2020.07.14.202119

[3] Goldstein N, Candau J and Clarke K 2004 *Computers, Environment and Urban Systems* **28** 125–147

[4] Lee J A 2020 Artificial neural networks in public policy: Towards an analytical framework

[5] Zhang C, Miao C, Zhang W and Chen X 2018 *Habitat International* **79** 51–60 ISSN 0197-3975 URL https://www.sciencedirect.com/science/article/pii/S0197397518301371

[6] Jaderberg M, Dalibard V, Osindero S, Czarnecki W M, Donahue J, Razavi A, Vinyals O, Green T, Dunning I, Simonyan K, Fernando C and Kavukcuoglu K 2017 *ArXiv* **abs/1711.09846**

[7] Dane. estimates of demographic change https://www.dane.gov.co/index.php/estadisticas-por-tema/demografia-y-poblacion/estimaciones-del-cambio-demografico accessed: 2021-08-30

[8] Data bank. world development indicators https://databank.bancomundial.org/reports.aspx?source=2&country=COL accessed: 2021-08-30

[9] Chen S, Chen B and Fath B 2014 *Environmental pollution* **190** 139–49

[10] Gounaridis D, Chorianopoulos I and Koukoulas S 2018 *Applied Geography* **90** 134–144 ISSN 0143-6228 URL https://www.sciencedirect.com/science/article/pii/S0143622817300498

IOP Conf. Series: Materials Science and Engineering 1253 (2022) 012008 doi:10.1088/1757-899X/1253/1/012008

Theoretical study of validation methodologies for energy simulation in nZEB buildings.

J G Ascanio-Villabona[1], O Lengerke-Pérez[1], C L Sandoval-Rodríguez A D Rincon-Quintero[1] and K T Jaimes- Quintero[1].

[1] Unidades Tecnológicas de Santander UTS, Bucaramanga Santander 680005, Colombia

Jgascanio@correo.uts.edu.co, Olenguerke@correo.uts.edu.co, Csandoval@correo.uts.edu.co Aricon@correo.uts.edu.co and Ktjaimes@uts.edu.co.

Abstract. Validation methodologies play an important role in the analysis of energy simulations in buildings [1], since it is used to reduce the uncertainty of the extrapolations that must be taken into account for any validation study, so there may be various methodologies [2]. This study focuses on developing a state of the art on validation methodologies that can be applied in the analysis of energy simulations for NZEB buildings. First, the review method compiles the documents relevant to the research selected by selection criteria, which indicate whether they are of scientific criteria. Secondly, the methodologies found are reviewed and described, in addition to discussing a series of studies applied to each of the methodologies. Finally, it can be concluded that the application of validation methodologies depends on the purpose of the research, however, Sensitivity analysis is the simplest and most useful method to use in the analysis of building performance, even if it has defects; due to its low computational price, easy execution and interpretation.

1. Introducción

Among the challenges for the XXI century is energy consumption and the impact for climate change[3]. Because the stock of buildings worldwide holds about 40% of total energy consumption in addition to producing a third of the total sum of CO2 emissions [4]. Therefore, to reduce these percentages it is essential to minimize energy consumption and achieve sustainability in buildings [5]. Currently in the construction industry according to the literature there is a difference between buildings in the projection stage and existing ones. According to the literature on energy consumption, for the first stage only a small total percentage belongs to it, however, with the integration of renewable energies and technological development, it is believed that within a few years not far from the current one these buildings will be of an independent energy nature. [3]. Otherwise for the second stage (existing buildings), since here belongs the majority of the total percentage so alternatives for improvement must be taken to reduce energy consumption.

The useful life cycle of a building ranges between 50 years or more, these buildings are subject to climates that as time progresses will be warmer and more volatile. [6]–[8]. For this reason it is important to modernize these buildings. To begin with, an analysis of the current energy consumption must be carried out, in order to plan the remodeling through refurbishment strategies either for commercial, residential, industrial and institutional buildings. [9]. Despite the above, this evaluation is not easy to work with due to the multiple variables that influence energy efficiency, so all these subsystems play a fundamental role that must be taken into account.

This is why energy simulation of buildings (BES) is an important tool not only in design but also to achieve energy certification in new buildings. In addition, these simulations are very useful to discover an anomaly, for example, in the use of energy resources, additionally alternatives can be evaluated that help optimize efficiency. However, the results that are generated are not accurate either for any method that is being used, so it has been a topic of discussion within the scientific community. [10]. Because, due to errors outside the acceptable margin can generate a negative effect on energy efficiency metrics underestimating a potential energy saving. In accordance with the above, scientists highlight the idea of having accurate but abbreviated models and thus generate realistic results[11].

Conventional models that are based on simulation use tools such as ECOTEC, EQUEST, ESP-r, TRNSYS or EnergyPlus [12]. These programs are considered today as the most used methods. That is why, the main objective of this article is to carry out a bibliographic review on validation methodologies through the compilation of studies through documentary sources such as Science Direct, Mendeley, Scopus, Google Scholar, among others[13]. In order to know tools that allow analyzing the results of energy simulations in buildings describe these methodologies found, in addition to mentioning the advantages and disadvantages of these.

2. Validation methodologies

Efficient buildings are a fundamental element in the alternatives and strategies of energy and sustainability advancement, representing a great techno-economic and socio-economic challenge. The reduction of CO2 emissions in buildings has become one of the main objectives of the policies, due to its consideration of the energy impact on buildings on a global scale. [14].

As time progresses, the construction sector will be influenced by mandatory standards, codes and norms in order to achieve near-zero energy in buildings (nZEB) [15], [16]. The European Building Performance Directive (EPBD) requires new buildings to be constructed as nZEB buildings by 2020, including existing constructions through renovation [17]. Despite the recent confinement generated by the COVID-19 virus [18] it is possible that related socio-economic problems delay this process.

For the dynamic simulation of buildings there are tools that are generally recognized as an appropriate method for accurate evaluation in the performance of buildings. So the development of nZEB standards, however, can still be observed more or less large discrepancies between simulated and real features for both new and existing buildings. [7], [19], [20]. In addition to the above, the real gap presented in several cases is too large to be acceptable; of which, the measured energy use could be up to 2.5 times the intended energy use [20], [21].

Therefore, there are validation methodologies to minimize the uncertainty of extrapolations that must be taken into account. Among these are:

2.1. Comparative studies

Comparative studies include direct comparisons of the results obtained from two or more simulations of the same initial data. This is a useful technique because it does not require real construction data. These models can be created and placed in real or fictional environments, so that researchers can suggest extreme situations and help understand the different heat transfer mechanisms that occur in buildings more quickly. [22].

In this way, the accuracy of the startup data can be fully controlled and external errors can be easily eliminated. On the other hand, comparative studies make it possible to simulate a large number of case studies in a relatively short period of time because they can be simplified to the maximum according to the needs of the real building. Case studies can be ordered to analyze the sensitivity of the simulation to various initial data changes.

The authors Zavrl & Stegnar [23], investigated energy use by calculating and measuring by monitoring an energy-efficient apartment building called Eco Silver House, which committed to implementing national nZEB building requirements.

The authors followed 20 apartments (some in full use, others unoccupied, but with moderate heating) by comparing two floors using the monthly method (PHPP and national) and with simulations in the IDA ICE software. Where they obtained as results the verification of the energy performance in the floors that were simulated to transient contour conditions, in addition to the fact that the building complied with energy parameters determining as an advantage this method of calculations applied in the comparison of the 2 floors.

2.2. *Sensitivity analysis*

It is defined as a sensitivity analysis: "Research on how to attribute the uncertainty of model results (numerical or other) to different sources of uncertainty of model input data."

Sensitivity analysis plays an important role in the energy research of buildings. It can be used to detect key variables that affect the thermal behavior of buildings, either in energy simulation models or under observational research. The sensitivity analysis method is the same in different types of building energy simulation applications and consists of several steps (See figure 1). The steps are:

1. Determine changes in input data;
1. Establish the research model;
2. Execute the energy model;
3. Collect the results of the simulation;
4. Run the sensitivity analysis; Present the results of the sensitivity analysis. It presents these steps to follow in the application [24].

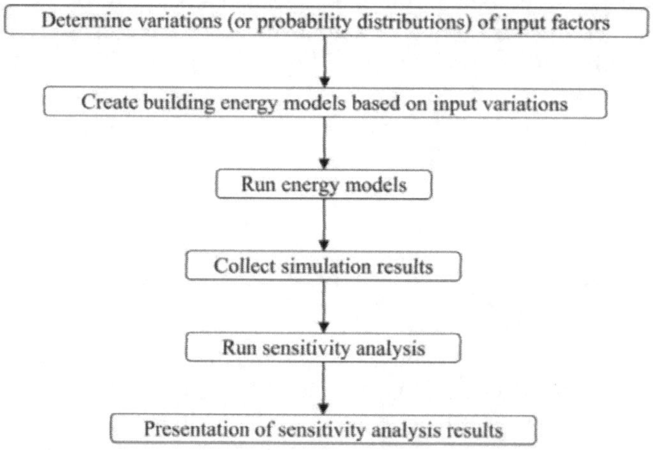

Figure 1. Steps to apply sensitivity analysis in energy simulation of buildings

Basically, sensitivity analysis measures how much relatively small changes in parameter values (input data) affect model results. It has a lot of utilities, [25]:

The authors Gelesz et al. [26] analyzed the sensitivity of double-skin façade models (DSFs) incorporated in two popularly known building energy software (BES) tools, EnergyPlus and IDA ICE, for various orientations and climates. Within the input variables, minimal changes were made in order to determine and identify the initial and fundamental parameters, with which those in charge of the design must take into account in the design process of the DSF according to the performance guidelines. Within the results they found that to determine the performance, the variables of the climate or orientation are not as relevant as the optical properties of the systems such as glazing and shading, followed by the thermal properties of the glazing. In addition, geometric characteristics, airflow and frame were not as important in this research. Finally, the validation process also revealed that differences in the models used lead to an inequality in the reliability of the BES tools used. From this point of view in a sensitivity analysis, it can be inferred that from one tool to another the results should be the same when defining the optimal solution for the system. Because IDA ICE facilitates the testing of more inputs than the EnergyPlus program, this sensitivity analysis exposes that this fact is not of great importance, since the most sensitive parameters are available in the two tools.

Likewise, the authors Zuhaib et al. [27] applied a methodology that would be used to generate calibrated and reliable models by means of automated methods in residential constructions no older than 35 years, using the values of the limited measurements and with high uncertainty. This multi-phase calibration method facilitates the combination of various types of data in step-by-step calibration. In addition, it allows to analyze not only models with variations of Indoor Environmental Quality (IEQ) but also analyzes it with the energy demand. Due to the large number of uncertainties in this model, a sensitivity analysis was performed. [28]; this model was calibrated according to Guideline 14 of ASHRAE and IMPVP where they present the indices of acceptable tolerances. (See Table 1)

Table 1. Acceptable tolerances for the validation of energy models of buildings

Calibration type	Index	ASHRAE Guide 14	IMPVP
Monthly	NMBE	5%	20%
	Cv (RMSE)	15%	-
Hourly	NMBE	10%	5%
	Cv (RMSE)	30%	20%

Note: NMBE: Normalized Mean Bias Error, Cv (RMSE): Coefficient of variation of the root of the mean square error. *Lower values indicate better calibration [29], [30].

Within the results, the phased approach methodology showed effectiveness in the creation of calibrated models in existing constructions with high uncertainty where they can be used as an influence on high-level decision-making in systems with a focus on improving the quality of the interior environment and energy performance.

2.3. Empirical verification

In empirical verification, the simulation results are compared with the data. Obtained from real buildings or test prototypes. Compare variables They present a certain degree of uncertainty due to errors in direct measurement. In addition, these errors will also result in uncertainty in the input data, which will lead to certain uncertainties when they spread throughout the simulation process. result. The total uncertainty corresponds to the combination of the two, and when measuring all the initial data, all deviations between the measurement result and the calculated result above this uncertainty threshold are attributed to the numerical solution or the component 5 and 6 or 7 error source model.

In this way, the authors Nageler et al. [31] applied a validated methodology in the design of fully automated constructions within urban areas supported by information available to the public. These models were created with heating systems in the IDA ICE simulation tool. The simulation was validated with the consumption data of the buildings supplied by the town of Gleisdorf (Austria). The results present an adequate approach to the measurement data, since a deviation of 0.98% is shown. this indicates that the method allows to model and describe the building stock of this locality, which means saving time facilitating the development for intelligent energy systems in the future.

2.4. *Quantification of uncertainty*

The proposed definition of uncertainty is: "Uncertainty to imply that, in a given situation, a person does not have the insufficient information, quantity, and quality to describe, specify, or predict a system, its behavior, or any other characteristic with certainty and numerically". According to Zimmermann [32] there are six causes of uncertainty:

- The lack of information requires the collection of additional information, but it should be considered that the quality and quantity of the information must be sufficient and adequate.
- Wealth of information (complexity)
- Contradictory evidence, sometimes the result of a model is contradictory with other previous results or evidence in situ, it is necessary to evaluate whether these contradictions are due to errors or real existence. Ambiguity, when information is provided in a format that can lead to confusion.
- Measurement uncertainty caused by lack of accuracy can be solved by the following methods More accurate measurement technology. However, investing in expensive equipment does not guarantee that there will be no errors.
- Previous judgments of researchers when evaluating model results may be different from those of other researchers.
- The judgments of the researcher may vary at the time of evaluating the results with those of another researcher.

2.4.1. *Uncertainty in the energy simulation of buildings.*

Energy simulation of buildings has become an indispensable support tool, but its reliability cannot be ignored. Evaluation is important Reliability and uncertainty of simulation and measurement to improve building design. Normally, to check the behavior of the building defined in the project phase, once the building is finished, a series of sensors are placed in the building. Compare them to the simulation results and you will usually find differences. Today, there are a lot of widely validated simulation tools, and these differences are usually due to errors in the introduction of parameters such as occupancy scenarios or climate data, such as the case study of a Won-jun building in Korea. [33].

On the other hand, the authors Jain et al. [34], used the methodology that is based on calibration in order to be able to identify and validate the main source of performance gaps. Following performance analysis in four case studies located in the UK. In the development of the calibration process and performance evaluation, the variables of Indoor Environmental Quality (IEQ) are used to optimize the reliability of said validation in the calibrated model and also highlight the interconnection between energy and IEQ. (See figure 2)

Figure 2. Methodology for the evaluation of the performance gap through calibrated energy simulation

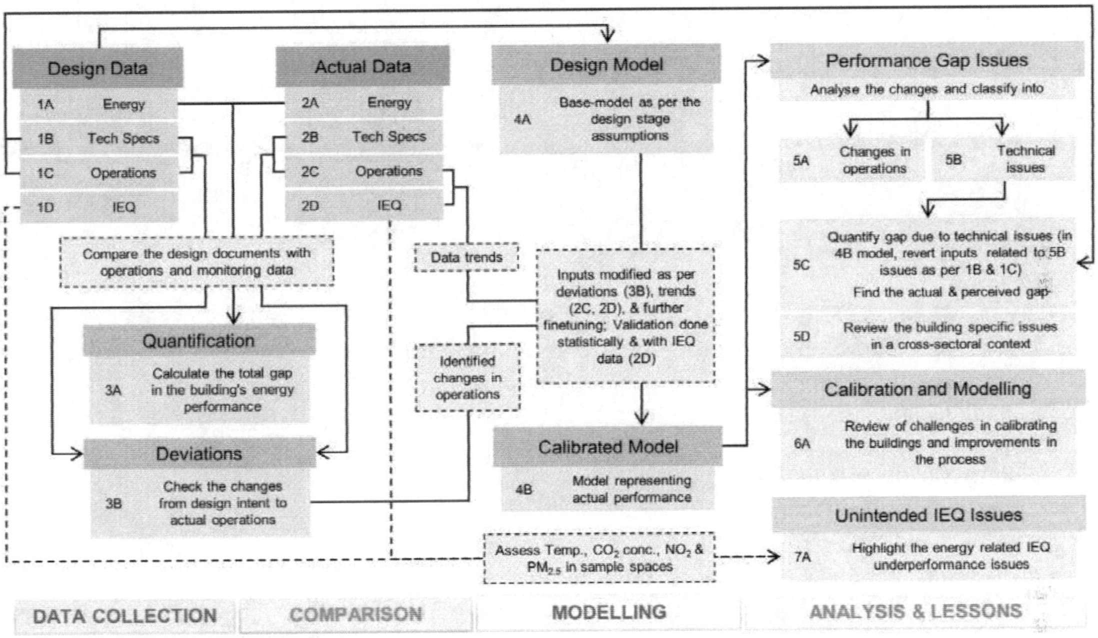

Among the results of the analysis, the analysis showed the relevance in the commitment to reduce the inconveniences with performance. In the relevant causes recognized in the case studies, they were presented according to three factors:

- Inadequate calculation methods in the design phase so it presents a gap observed by the use of this method. This problem could be solved by optimizing industry practices.
- Technical drawbacks in the construction, systems used and operations in the building.
- The last factor is due to the operational variations for which the building has been exposed in order to meet the functional requirements. This factor is due to the variation caused by necessity; gaps caused by technical problems need to be observed and corrected to ensure that the building works optimally.

On the other hand, the authors Tardioli et al. [35] with the non-deterministic calibration method applied to a set of buildings they used four techniques (See figure 3). In the first, groupings are used where the set of buildings and distinctive constructions grouped to the urban area can be identified. In the second technique, they applied a focus on energy modeling in order to perform simulations on these buildings. For the third, they apply the models based on data used as emulators of the dynamic simulation engine, in this way it is developed computationally and efficiently. Finally, the fourth technique consists of the exploitation of a calibration frame with the non-deterministic Bayesian method with the aim of determining the fundamental parameters for the group of buildings.

Figure 3. Overview of research methodology in terms of workflow

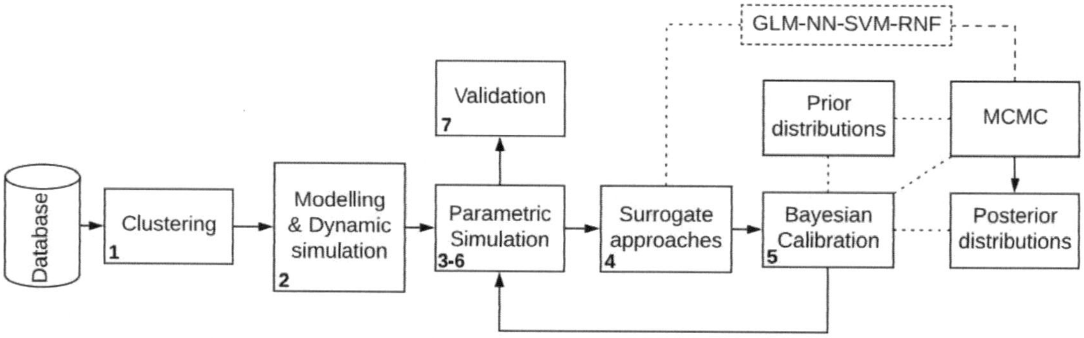

The results show a distribution of modeling parameters characterized by groups of construction models. This validation with the data obtained in the measurement, represent the energy consumption and the predictions in the energy intensity.

2.5. Analytical verification

A common building energy simulation tool includes various variables and parameters. The number of case studies that can be simulated by combining all of these is astronomical, so they can never be completed in their entirety. However, all case studies are governed by the same basic heat transfer mechanisms: conduction, radiation, and convection. Therefore, a simple enough analysis case study can be defined, in addition to simulation, it can also be solved through analysis [36].

These case studies will be better than real systems, but they can be defined as basic heat transfer mechanisms that can have a greater impact on their thermal behavior in isolation. In this way, most of the important content of the simulated pyrolysis algorithm can be easily identified and isolated.

3. Advantages and disadvantages of validation methodologies

Finally, the advantages and disadvantages of different validation techniques described above are presented (See figure 4)

Figure 4. Advantages and disadvantages of different validation techniques.

Technique	Advantages	Disadvantages
Comparative Relative test of model and solution process	- No input uncertainty - Any level of complexity - Inexpensive - Quick: many comparisons possible	No truth standard
Analytical Test of numerical solution	- No input uncertainty - Exact truth standard given the simplicity of the model - Inexpensive	- No test of model - Limited to cases for which analytical solutions can be derived
Empirical Test of model and solution process	- Approximate truth standard within accuracy of measurements - Any level of complexity	- Measurement involves some degree of input uncertainty - Detailed measurements of high quality are expensive and time consuming - A limited number of data sites are economically practical

4. Conclusions

This project has reviewed the validation methodologies for energy simulations of NZEB buildings. The findings and recommendations of the application of these methods and techniques to evaluate thermal performance in buildings are summarized below.

First, the work on the simulation shows that the computer model can faithfully reproduce the complexity and difficulty of the building's actual behavior. The most faithful calibration possible of the model is understood as a basic step, so it is possible to predict the behavior of the future model or the change of the existing model.

The construction model validation process is an iterative process of comparing simulation results with results obtained from measurement data to determine the deviation and uncertainty of the results. The process begins with the configuration of a numerical model created as accurately as possible, assigning all the information about the geometry of the building, thermal characteristics, actual meteorological data, the characteristics of the building's electrical and air conditioning systems, control logic and formulating equipment, activities and occupational schedules.

The disadvantages of analytical methodology is that it exists in a limited number of configurations from which analytical solutions can be derived. In addition, analytical verification only allows to check the adequacy of the numerical part of the internal error, not the validity of the model itself. Although it is a simulation When applied to a real system, it may not be correct in all analysis checks. When combined with comparative research, it increases the potential for analytical verification. From a very simple analysis case study, it is possible to establish reference correlation between different analytical and simulation solutions. From here, you can gradually modify the parameters of the analysis case study to better respond to the complexity of the actual system.

Sensitivity analysis is the simplest and most useful method to use in the analysis of building performance, even if it has defects; due to its low computational price, easy execution and interpretation. However, this method has limitations since it only digitizes the space in the input data. Modifications to the analysis input depend on the objective of the study. If the researcher plans to investigate the impacts on the different design alternatives, the arrangement of the variables can use a uniform distribution. On the other hand, if the intention is only to study the level of thermal performance presented by existing constructions, normal distributions are used for most of the variables.

IOP Conf. Series: Materials Science and Engineering 1253 (2022) 012008 doi:10.1088/1757-899X/1253/1/012008

References

[1] L. Evangelisti, C. Guattari, F. Asdrubali, and R. de Lieto Vollaro, "In situ thermal characterization of existing buildings aiming at NZEB standard: A methodological approach,"
Dev. Built Environ., vol. 2, no. March, p. 100008, May 2020, doi: 10.1016/j.dibe.2020.100008.

[2] H. E. Huerto-Cardenas *et al.*, "Validation of dynamic hygrothermal simulation models for historical buildings: State of the art, research challenges and recommendations," *Build. Environ.*, vol. 180, no. March, p. 107081, 2020, doi: 10.1016/j.buildenv.2020.107081.

[3] E. Asadi, M. G. Da Silva, C. H. Antunes, and L. Dias, "Multi-objective optimization for building retrofit strategies: A model and an application," *Energy Build.*, vol. 44, no. 1, pp. 81– 87, Jan. 2016, doi: 10.1016/j.enbuild.2011.10.016.

[4] R. Ruparathna, K. Hewage, and R. Sadiq, "Improving the energy efficiency of the existing building stock: A critical review of commercial and institutional buildings," *Renewable and Sustainable Energy Reviews*, vol. 53. Elsevier Ltd, pp. 1032–1045, Jan. 2016, doi: 10.1016/j.rser.2015.09.084.

[5] M. Karmellos, A. Kiprakis, and G. Mavrotas, "A multi-objective approach for optimal prioritization of energy efficiency measures in buildings: Model, software and case studies," *Appl. Energy*, vol. 139, pp. 131–150, Feb. 2015, doi: 10.1016/j.apenergy.2014.11.023.

[6] Y. H. Yau and S. Hasbi, "A review of climate change impacts on commercial buildings and their technical services in the tropics," *Renewable and Sustainable Energy Reviews*, vol. 18. Pergamon, pp. 430–441, Feb. 2018, doi: 10.1016/j.rser.2012.10.035.

[7] P. de Wilde and D. Coley, "The implications of a changing climate for buildings," *Building and Environment*, vol. 55. Pergamon, pp. 1–7, Sep. 2012, doi: 10.1016/j.buildenv.2012.03.014.

[8] M. Olonscheck, A. Holsten, and J. P. Kropp, "Heating and cooling energy demand and related emissions of the German residential building stock under climate change," *Energy Policy*, vol.
39, no. 9, pp. 4795–4806, Sep. 2016, doi: 10.1016/j.enpol.2011.06.041.

[9] Y. Fan and X. Xia, "Energy-efficiency building retrofit planning for green building compliance," *Build. Environ.*, vol. 136, pp. 312–321, May 2018, doi: 10.1016/j.buildenv.2018.03.044.

[10] M. Herrera *et al.*, "A review of current and future weather data for building simulation," *Build.*
Serv. Eng. Res. Technol., vol. 38, no. 5, pp. 602–627, Sep. 2017, doi: 10.1177/0143624417705937.

[11] J. G. Ascanio-Villabona, C. L. Sandoval-Rodriguez, A. D. Rincón-Quintero, B. E. TarazonaRomero, and R. E. Paez-Castro, "Building a prototype for functional analysis of the energy potential of the water flow in pipe ½ "using microturbines applied to Unidades Tecnológicas de Santander," *IOP Conf. Ser. Mater. Sci. Eng.*, vol. 844, no. 1, p. 012056, May 2020, doi:
10.1088/1757-899X/844/1/012056.

[12] V. M. Zavala, "Real-time optimization strategies for building systems†," *Ind. Eng. Chem. Res.*, vol. 52, no. 9, pp. 3137–3150, Mar. 2017, doi: 10.1021/ie3008727.

[13] J. G. Ascanio-Villabona, L. A. Del Portillo-Valdés, O. Lengerke-Pérez, B. E. T. Romero, A. D. Rincón-Quintero, and M. A. Durán-Sarmiento, "Analysis of the Energy Potential of a Tangential Microturbine for Application in a Passivhaus Environment," Springer, Cham, 2021, pp. 181–195.

[14] U. Berardi, "A cross-country comparison of the building energy consumptions and their trends," *Resour. Conserv. Recycl.*, vol. 123, pp. 230–241, Aug. 2017, doi:

10.1016/j.resconrec.2016.03.014.
[15] J. RE, W. TE, B. LW, D. T, C. E, and Colliver DG, "ASHRAE 2020 Ad Hoc committee committee Vice-Chair, McQuay international Mark S. Menzer, air conditioning, heating and refrigeration institute," 2020.
[16] E. Parliament., "European parliament report on the proposal for a directive of the European parliament and of the council on the energy performance of buildings (recast) (COM (2008)0780-C6-0413/2008-2008/0223(COD)), 2009.," 2019.
[17] B. Conterio, "Directiva 2010/31 / UE sobre la eficiencia energética de los edificios (refundición) - 19 de mayo de 2010 | Construir," 2020. .
[18] J. Parada, "Comprensión del impacto sectorial de COVID-19: ingeniería y construcción," 2019.
.
[19] N. Kampelis *et al.*, "Evaluation of the performance gap in industrial, residential & tertiary nearZero energy buildings," *Energy Build.*, vol. 148, pp. 58–73, Aug. 2017, doi: 10.1016/j.enbuild.2017.03.057.
[20] A. C. Menezes, A. Cripps, D. Bouchlaghem, and R. Buswell, "Predicted vs. actual energy performance of non-domestic buildings: Using post-occupancy evaluation data to reduce the performance gap," *Appl. Energy*, vol. 97, pp. 355–364, Sep. 2018, doi: 10.1016/j.apenergy.2011.11.075.
[21] P. De Wilde, "The gap between predicted and measured energy performance of buildings: A framework for investigation," *Autom. Constr.*, vol. 41, pp. 40–49, May 2020, doi: 10.1016/j.autcon.2014.02.009.
[22] A. G. Muñoz, "Validación y calibración de la simulación energética de edificios. La importancia del análisis de sensibilidad e incertidumbre," 2018.
[23] M. Š. Zavrl and G. Stegnar, "Comparison of Simulated and Monitored Energy Performance Indicators on NZEB Case Study Eco Silver House," *Procedia Environ. Sci.*, vol. 38, pp. 52–59,
2017, doi: 10.1016/j.proenv.2017.03.074.
[24] Tian and Wei, "A review of sensitivity analysis methods in building energy analysis," *Renew. Sustain. Energy Rev.*, vol. 20, no. C, pp. 411–419, 2019.
[25] M. Eriksson, J. Akander, and B. Moshfegh, "Development and validation of energy signature method – Case study on a multi-family building in Sweden before and after deep renovation," *Energy Build.*, vol. 210, p. 109756, 2020, doi: 10.1016/j.enbuild.2020.109756.
[26] A. Gelesz, E. Catto Lucchino, F. Goia, V. Serra, and A. Reith, "Characteristics that matter in a climate façade: A sensitivity analysis with building energy simulation tools," *Energy Build.*, vol. 229, p. 110467, 2020, doi: 10.1016/j.enbuild.2020.110467.
[27] S. Zuhaib, M. Hajdukiewicz, and J. Goggins, "Application of a staged automated calibration methodology to a partially-retrofitted university building energy model," *J. Build. Eng.*, vol. 26, no. July, p. 100866, 2019, doi: 10.1016/j.jobe.2019.100866.
[28] F. Roberti, U. F. Oberegger, and A. Gasparella, "Calibrating historic building energy models to hourly indoor air and surface temperatures: Methodology and case study," *Energy Build.*, vol.
108, pp. 236–243, Dec. 2018, doi: 10.1016/j.enbuild.2015.09.010.
[29] Committee Ipmvp, "International Performance Measurement & Verification Protocol; Volume 1: Concepts and Options for Determining Energy and Water Savings," 2016.
[30] ASHRAE, "Measurement of Energy, Demand, and Water Savings," Atlanta, 2014.
[31] P. Nageler *et al.*, "Novel validated method for GIS based automated dynamic urban building energy simulations," *Energy*, vol. 139, pp. 142–154, 2017, doi: 10.1016/j.energy.2017.07.151.

[32] Jorge Veiga de Cabo, Elena de la Fuente Díez, and Marta Zimmermann Verdejo, "Modelos de estudios en investigación aplicada: conceptos y criterios para el diseño," *Med. Segur. Trab. (Madr).*, vol. 54, no. 210, pp. 81–88, 2018.
[33] W.-J. Suh, C.-S. Park, and D.-W. Kim, "APPLICATION OF A WHOLE BUILDING SIMULATION TOOL FOR A REAL-LIFE BUILDING," in *12th Conference of International Building Performance Simulation Association*, 2019, pp. 1–8.
[34] N. Jain, E. Burman, S. Stamp, D. Mumovic, and M. Davies, "Cross-sectoral assessment of the performance gap using calibrated building energy performance simulation," *Energy Build.*, vol. 224, p. 110271, 2020, doi: 10.1016/j.enbuild.2020.110271.
[35] G. Tardioli, A. Narayan, R. Kerrigan, M. Oates, J. O'Donnell, and D. P. Finn, "A methodology for calibration of building energy models at district scale using clustering and surrogate techniques," *Energy Build.*, vol. 226, p. 110309, 2020, doi: 10.1016/j.enbuild.2020.110309.
[36] H. Jia and A. Chong, "eplusr: A framework for integrating building energy simulation and data-driven analytics," *Energy Build.*, vol. 237, p. 110757, Apr. 2021, doi: 10.1016/j.enbuild.2021.110757.

IOP Conf. Series: Materials Science and Engineering 1253 (2022) 012009 doi:10.1088/1757-899X/1253/1/012009

Exergy analysis and development of flat plate solar collectors: A Review

A D Rincón-Quintero[1,2], L A Del Portillo-Valdés[2], N D Zanabria-Ortigoza[1], C L Sandoval-Rodriguez[1,3], J G Maradey-Lázaro[4] and N Y Castillo-León[1]

[1] Design and materials research group (DIMAT), Automation and Control Energy Systems Research Group (GISEAC), Electromechanical engineering, Faculty of Natural Sciences and Engineering, Unidades Tecnológicas de Santander UTS, Calle de los estudiantes No 9-82, 680005, Bucaramanga, Colombia.
[2]Energy in Building research group (ENEDI), Doctoral Program in Energy Efficiency and Sustainability in Engineering and Architecture, Department of Energy Engineering, University of the Basque Country, UPV/EHU, Engineer Torres Quevedo Plaza, 1, 48013, Bilbao, Spain.
[3]Doctoral Program in Electronic and Communications, Department of Communications Engineering, University of the Basque Country, UPV/EHU, 48013, Bilbao, Spain.
[4]Doctoral program in engineering, Universidad Autónoma de Bucaramanga UNAB, Carrera. 47 #53-78, 680003, Bucaramanga, Colombia.

arincon@correo.uts.edu.co

Abstract. The exergy comes from the need to find some energy of a system that can be quantified to take advantage of its mechanical, electrical or any other type of work, it is limited in terms of investment of time and research, due to this fact, the need arises to be able to innovate with data collection on the exergy calculation models developed in flat solar collectors. In the present research, a search of the literature regarding the development of exergy in flat plate solar collectors was carried out, to then analyze and determine forty innovative articles in the last six years, which were reflected in a data matrix in the The formulas and factors that define the exergy are presented, as well as the theoretical designs of the collector and working fluids used in flat solar collectors. With the results of the table, a broad panorama of current data is obtained, on the most used in the formulation of exergetic efficiency, which stands out as an essential tool for the design, analysis and optimization of thermal systems.

Index Terms: Exergy, Flat plate solar, Solar collectors, Optimization of thermal systems, Working fluids.

1. Introduction

In recent years, the need to obtain resources for renewable energies, such as solar energy, has increased considerably, since it is friendly to the environment, freely accessible when using it in any labor field and important in the development of many technologies in history. The flat solar collector is an adaptation that is currently used to supply domestic and commercial networks with thermal energy, mainly in heating or cooling applications in specific environments.

n the application that is developed. Also, it is very important to contribute this study to subsequent projects in the same framework, so that they have current antecedents of what was done, and thus improve even more, the exergy obtained in the model that is being sought.

The objective of this article is to be based on the bibliographic analysis of the literature referring to the study of energy in flat solar collectors, and then to make a selection of forty most outstanding scientific articles in the last six years. The authors of the different sections focus mainly on the design of the flat plate solar collector, working fluid and the exergetic efficiency of the collector, which allows to analyze and quantify the useful working capacity of the system with its environment.

This review presents a methodological process that consists of the bibliographic analysis and the creation of a data matrix focused on the technological models already investigated, in order to summarize and visualize the analysis made. The table will show the thermodynamic factors and the necessary formulas to determine the exergetic efficiency, used by the authors of the investigations, emphasizing the effectiveness in each one of them. Also, the design of the flat plate solar collector that was based on the energy analysis previously exposed in each model will be defined. Then the working fluid used to deliver energy to the heating will be exposed, after being heated by the flat solar collector. Finally, an analysis of the results will be carried out with the creation of graphs to extract the information from the previous table, and thus be able to visualize the investigated models in an easier way.

2. Work development

2.1 Bibliographic analysis

To begin the intensive search for articles regarding the development of exergy in flat plate solar collectors, the ScienceDirect database is approached, which allows easy and fast access to databases with a large number of scientific journals from high impact and world class.

In the first instance, it was decided to graph using VOSviewer the co-occurrence of keywords in articles that report on exergy in flat solar collectors, since this allows a brief idea about the content of the articles written in the last six years.

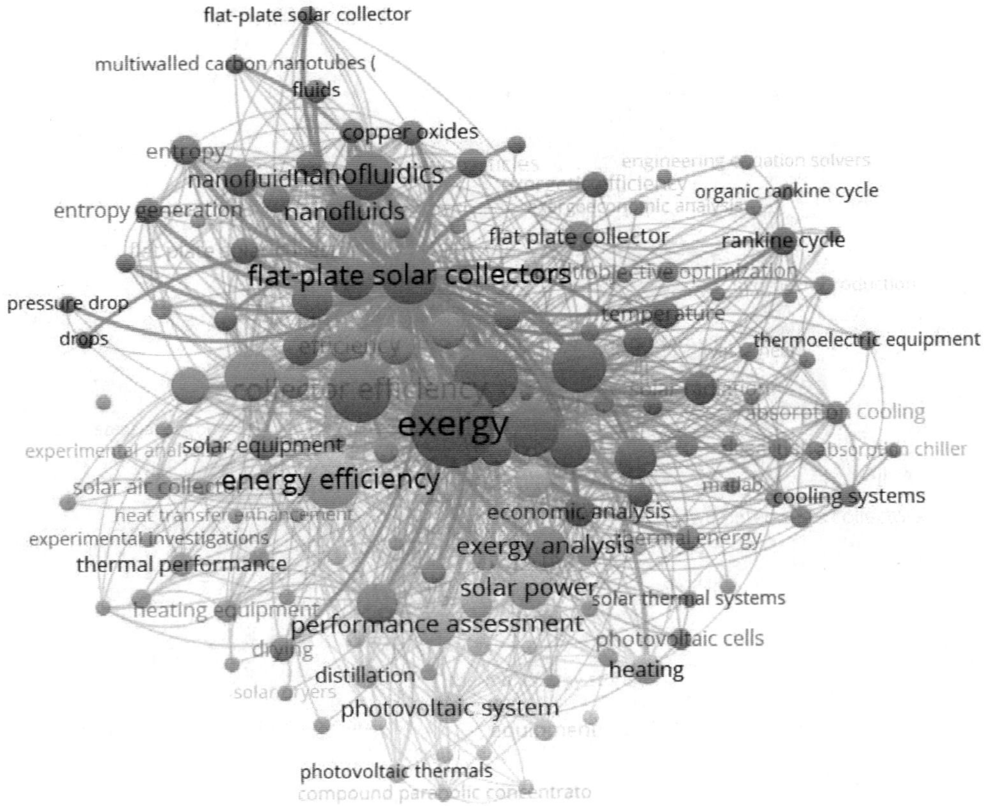

Figure *1*. Map of co-occurrences with key words highlighted in the selected technical-scientific literature. Source: Authors.

In Figure 1, a network of 144 nodes or keywords can be observed, which have at least 5 coincidences within 190 articles, in which it is possible to show the content of the studies in the different sections. In the network, three general lines of research are established, the first and broadest one focuses on exergy, linking topics such as rankine cycle, refrigeration systems and flat plate collector, deducing their applications in heating or cooling of specific environments. The second line of research is oriented to flat plate solar collectors, nanofluids and pressure drops, highlighting the collector design and the experimental use of different nanofluids in water to increase its efficiency. The third line of research focuses on energy efficiency, exercise analysis and solar energy. The latter is the one of greatest interest in this article since it allows quantifying the energy analysis of the efficient use of solar energy until reaching equilibrium with a reference environment.

To increase the processing of the registration of the selected scientific articles in the database, it is determined to graph in VOSviewer, a bibliographic coupling network that allows access with your selection to each document as shown below.

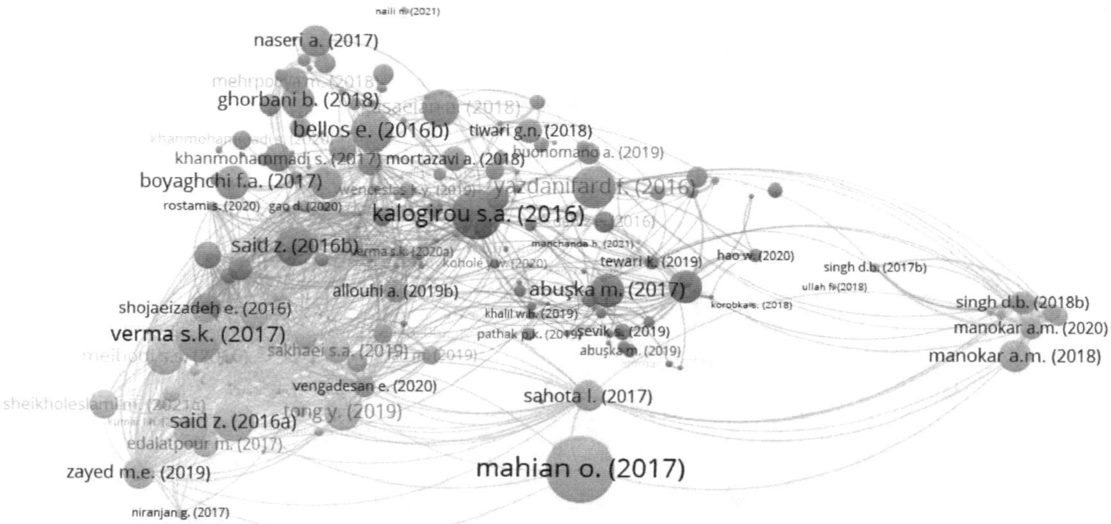

Figure 2. Bibliographic coupling network map highlighted in the selected technical-scientific literature. Source: Authors.

In Figure 2, a network of 190 nodes or documents can be observed, which have at least 2 coincidences within 190 articles, in which it is possible to show the bibliography of the different sections; With this network it is possible to identify those articles that have a significant impact through the links generated between them, those that have the greatest number of connections between other nodes within the network, are selected for reading in an orderly manner.

2.2 Data matrix

To detail the research of each of the forty technical-scientific articles, the thermodynamic factors used in the models and the formulas that allow to quantify the exergy as the efficiency of solar energy when applied are illustrated through a data matrix. to flat plate collectors, highlighting the effectiveness of each item investigated. Also, the design of the flat plate solar collector will be defined, which was based on the energy analysis previously exposed by the authors of the investigated projects. Finally, the working fluid used in each flat solar collector for use in heating specific places will be known. Next, it is presented in Table 1 of what was said above.

Table 1. Data matrix on the technical-scientific articles investigated.

Articles	Thermodynamic factors	Exergetic formulation	Solar collector design	Work flow	Exergy efficiency
[1]	-Destruction of exergy. -Ambient and solar temperature -Solar energy absorbed by collector.	$\eta_{ex} = 1 - \frac{\dot{E}x_{dest}}{\left[1-\left(\frac{T_a}{T_{sol}}\right)\right]\dot{Q}_{sol}}$ Equation 1.	-Area: 2.03 m^2. -Empty weight: 35.5 kg. -Fluid capacity 1.5L. -Absorbent material Copper with a thickness of 2 mm.	Water	21,9%
				Water with nanoparticles of Al_2O_3	59.6%
				Water with nanoparticles of CuO	49.6%

[2]	- Solar temperature. -Entropy generation rate. - Incident energy in the collector area.	$\eta_{ex} = 1 - \frac{T_e \dot{S}_{gen}}{\left[1-\left(\frac{T_e}{T_s}\right)\right]\dot{Q}_s}$ Equation 2.	- Steel sheet: 1 × 2 × 0.2 m -Transparent layer of tempered glass. -Absorbent material Copper or Aluminum with thickness of 0.2 and 0.1 mm.	Air	6 a 12%
[3]	- Apparent solar and ambient temperature. -Exergy rate. -Global solar radiation. -Surface of the solar collector.	$\eta_{ex} = \frac{E\dot{x}_{in} - E\dot{x}_{in,f}}{G_T A_c\left[1-\frac{T_a}{T_s}\right]}$ Equation 3.	- Area: 1.51 m^2. -Weight: 38.5 kg. -Diameter of the collector tubes 8.3 mm. -Collecting plate with a thickness of 5 mm.	Water with nanoparticles of Al_2O_3	4.3 a 12.53%
[4]	- Ambient and solar temperature. -Entropy generation rate. -Energy absorbed by the collector.	$\eta_{ex} = 1 - \frac{T_a \dot{S}_{gen}}{\left[1-\left(\frac{T_a}{T_s}\right)\right] q_{abs}}$ Equation 2.	- Absorption area: 2 m^2. -Fluid capacity: 3 L. -7 risers. -Absorbent material Copper.	Distilled water with multi-walled carbon nanotubes (MWCNT).	23.35%
[5]	- Solar temperature. -Entropy generation rate. - Incident energy in the collector area.	$\eta_{ex} = 1 - \frac{T_e \dot{S}_{gen}}{\left[1-\left(\frac{T_e}{T_s}\right)\right]\dot{Q}_s}$ Equation 2.	- Absorption area of 0.375 m^2. -Edge area 0.16 m^2 -Weight: 15 kg. -Glazing thickness 4 mm. -Emissivity of absorbent plate 0.12.	Water	4.62%
				Distilled water with multi-walled carbon nanotubes (MWCNT).	29.32%
				Water with graphene nanoparticles.	21.46%
				Water with nanoparticles of CuO.	16.67%
				Water with nanoparticles of Al_2O_3.	10.86%
				Water with nanoparticles of TiO_2	6.97%
				Water with nanoparticles of SiO_2	5.74%
[6]	- Exergy rate gained. -Collector area. -Incident solar radiation. -Ambient and solar temperature.	$\varepsilon_{ex} = 1 - \frac{\dot{E}_x}{A_s\left[I_s\left(1-\frac{T_a}{T_s}\right)\right]}$ Equation 4.	- Dimensions: 1941x1027x88 mm. -Glass cover with thickness of 3.2 mm. -Copper shock absorber and tubes.	Water with nanoparticles of Al_2O_3	3 a 7.7%
[7]	- Heat energy rate. -Surface of the collector. -Normal solar radiation to the collector.	$\eta_{ex} = \frac{Q_u}{A_c G_T}$ Equation 5.	- Dimensions: 1009x2009x75 mm. -Liquid capacity of 1.57 L. -Insulation material: 40 mm rock wool.	Deionized water with nanoparticles of CeO_2.	10.74%
[8]	- Heat energy rate. -Surface of the collector. -normal solar radiation to the collector.	$\eta_{ex} = \frac{Q_u}{A_c G_T}$ Equation 5.	- Dimension: 1009x2009x75 mm. -Liquid capacity of 1.57 L. -Insulation material: 40 mm rock wool.	Deionized water with nanoparticles of WO_3.	13.48%

[9]	- Exergy rate of absorbed solar radiation. -Absorbent plate area. -Incident solar radiation. -Ambient and solar temperature.	$\eta_{ex} = 1 - \frac{\dot{E}_{in,Q}}{I_T A_p \left(1 - \frac{T_a}{T_s}\right)}$ Equation 4.	- Dimension: 2000x1000x80 mm. -Weight: 36 kg. -Absorption area: 1.84 m^2. -Copper header and tubes.	Water with nanoparticles of Al_2O_3.	20.3%
[10]	- Ambient and solar temperature. -Entropy generation rate. - Incident energy in the collector area.	$\eta_{ex} = 1 - \frac{T_a \dot{S}_{gen}}{\left[1 - \left(\frac{T_a}{T_s}\right)\right] \dot{Q}_s}$ Equation 2.	- Absorption area: 1.84 m^2. -4 mm tempered glass. - Header and copper tubes TP2. -Emissivity of 0.12.	Water with nanoparticles of Al_2O_3.	20.3%
[11]	- Destruction of exergy. -Exnergy of the source.	$\eta_{ex} = 1 - \frac{Ex_d}{Ex_{source}}$ Equation 6.	- Absorber length and width: 1 and 1,092 m. -Internal diameter of tube: 0.017 m. -12 flow channels. -Absorbent plate and copper tube.	Water with nanoparticles of Fe_3O_4.	12.8%
[12]	- Ambient and solar temperature. -Entropy generation rate. - Incident energy in the collector area.	$\eta_{ex} = 1 - \frac{T_a \dot{S}_{gen}}{\left[1 - \left(\frac{T_a}{T_s}\right)\right] \dot{Q}_s}$ Equation 2.	- Dimension: 2300x1300x100 mm. -Area: 3 m^2. -Absorbent plate thickness: 0.6 mm. -Copper fin, head and riser.	Water with nanoparticles of nanodiamonds (ND)	3.49%
[13]	- Entropy generation rate. -Ambient and solar temperature. - Incident energy in the collector area.	$\eta_{ex} = 1 - \frac{\dot{S}_{gen} T_a}{\left[1 - \left(\frac{T_a}{T_s}\right)\right] \dot{Q}_s}$ Equation 2.	- Collector area: 1.135x0.6 m^2. -Area of the copper absorbent plate: 0.9144x0.508 m^2. -Four vertical tubes.	Distilled water with green covalent functionalized graphene nanoparticles.	24.09%
[14]	- Contributing exergy rate. -Production exergy rate.	$\varepsilon_{coll} = \frac{\dot{E}x_{out}}{\dot{E}x_{in}}$ Equation 7.	- Area: 1x3x0.1m (2 units) / 6 m^2. -Thickness of steel sheet of 0.001m. -Fabric arranged diagonally along the collector as an absorbent plate.	Air	41%
[15]	- Exergy flow rate. -Hydrogen production rate. -Lower calorific power. -Heat transfer rate.	$\eta_{ex} = \frac{\dot{E}x_{H2}}{\dot{E}x_{in,sun}}$ $\eta_{ex} = \frac{\dot{m}_{H2} LHV_{H2}}{\dot{Q}_{in,sun}}$ Equation 7.	The process consists of the flat plate solar collector, organic Rankine cycle and polymer exchange membrane electrolyzer.	Methanol.	48.69 a 54.96%
[16]	- Destruction of exergy. -Ambient and solar temperature.	$\eta_{ex} = \left[1 - \frac{(-\dot{E}_d)}{\left[1 - \left(\frac{T_a}{T_s}\right)\right] I A_p}\right]$	- Collector area: 2 m^2. -Absorbent plate area. 1.6 m^2.	Water.	8.3 a 11.3%

	-Intensity of radiation. -Absorbent plate area. -Pressure drop.	$-\frac{m\Delta P}{\rho}$ Equation 8.	-Simple glass with a thickness of 4 mm. -Insulation thickness: 70 mm.		
[17]	- Daily useful potential. -Instant useful thermal power. -Temperatures found in mathematical models.	$\psi_{ex} = \frac{\int_{t_{sr}}^{t_{ss}} \dot{Q}_u dt}{\int_{t_{sr}}^{t_{ss}} \dot{Q}_s dt}$ Equation 9.	- gross area of 2.27 m^2. -Comprises 20 copper heat pipes. -Effective irradiation area of 1,149 m^2. -Mass water flow rate 0.0458 kg/s.	Distilled water.	7.3%
[18]	- Exergy flow rate. -Exergy destruction rate. -Heat transfer rate.	$\eta_{ex} = \frac{\Sigma \dot{E}_{toss} + \Sigma \dot{E}_{dst}}{\dot{E}x_{in,sun}}$ Equation 10.	- Air enters at ambient temperature and pressure. -The glazing has no depth. -The absorbent plate is supposed to be thin.	Air	12.36 a 22.84%
[19]	- Useful production. -Input exergy rate.	$\eta_{ex} = \frac{E_u}{E_{in}}$ Equation 7.	- Collecting surface 0.8x1.6 m^2. -Clear polycarbonate cover, -Absorbent copper plate.	Air	7%
[20]	- Exergy rate of the desired product. -Exergy rate of the applied sources.	$\eta_{ex} = \frac{\dot{X}_{wanted}}{\dot{X}_{source}}$ Equation 7.	- Useful area of 4000 m^2 -Collector assisted at night, by an auxiliary heater, they provided thermal energy of 2627 kW to the Kalina energy cycle and the biogas.	Water	73%
[21]	- Exergy rate of absorbed solar radiation. -Ambient and solar temperature. -Surface of the collector. -Normal solar radiation to the collector.	$\eta_{ex} = \frac{\dot{E}_{in,Q}}{G_t A_c \left(1 - \frac{T_a}{T_s}\right)}$ Equation 3.	- A three-dimensional slanted sheet tube flat plate solar collector is numerically modeled. -Gross area: 1500x60x50 mm^2.	Water with nanoparticles of Al_2O_3.	10.96%
[22]	- Absorbed Exergy. -Exergy lost. -Exergy of entry of the sun to the collector.	$\eta_{ex} = \frac{Ex_{c2} - Ex_{c1}}{Ex_{sun,in}}$ Equation 11.	- 120 collectors were used. Length and width of each collector is 2x1.2 m. -Total aperture is 288 m^2.	Pentane	24.08%
[23]	- Exit Fluid Exergy. -Exergy of the inlet fluid. -Exergy absorbed by the solar collector.	$\eta_{ex} = \frac{Ex_{o,f} - Ex_{i,f}}{Ex_s}$ Equation 11.	-Dimensions: 2015x975x51mm (1.96 m^2). -Weight: 53 kg. -Capacity: $1.4x10^{-3}m^2$	Water mixed with propylene glycol antifreeze.	14.35%

[24]	- Exergy of entry and exit of the working fluid in the collector. -Exergy absorbed by the solar collector.	$\eta_{ex} = \frac{\Delta Ex_F}{Ex_{s,I}}$ Equation 11.	- Operating temperature range: 30–80 ° C. -Area: 8.3 m^2. -Tracking: Stationary. -Concentration ratio: 1	Multi-walled carbon nanotube water (MWCNT).	17.18%
[25]	- Exergy obtained by the receptor of the cavity. -Total incoming solar exergy.	$\eta_{ex} = \frac{\Sigma \dot{E}_{gained}}{\Sigma \dot{E}_{in,rec}}$ Equation 7.	- Longitudinal hub distance: 1.8 cm. -Cavity receiver opening diameter: 14 cm. -Height of the cavity receiver: 14 cm.	Thermal oil with nanoparticles of Al_2O_3.	11.732%
				Thermal oil with nanoparticles of SiO_2	11.73%
				Thermal oil with nanoparticles of TiO_2	11.738%
				Thermal oil with nanoparticles of Cu	11.75%
[26]	- Useful exergy output. -Input of solar exergy.	$\eta_{ex} = \frac{E_u}{E_{in}}$ Equation 7.	- Longitudinal distance of the concentrator: 3.8 m. -Collector made of AISI 304 stainless steel -It is added: absorber in a spiral shape, inside the casing.	Water	21.42%
[27]	- Entropy generation rate. -Ambient and solar temperature. - Incident energy in collector area.	$\eta_{ex} = 1 - \frac{\dot{S}_{gen} T_a}{\left[1 - \left(\frac{T_a}{T_s}\right)\right] \dot{Q}_s}$ Equation 2.	- Length: 1,135 m. -Width: 0.6 m. -Thickness: 0.09 m. -Absorbent area: 0.4645 m^2.	Water	25%
				Water with nanoparticles of SiO_2.	30%
				Water with nanoparticles of ZnO.	35%
[28]	- Useful exergy output. -The exergy input to the collector.	$\eta_{ex} = \frac{E_u}{E_{solar}}$ Equation 7.	- Dimensions: 3.4x2.46x0.66 m. -Area: 8,365 m^2. -Absorber area: 8.36 "m" ^ "2". -Volume: 5,520 m^3.	Air	24%
[29]	- Useful exercises in the collector. -Exergy absorbed from solar rays.	$\eta_{ex} = \frac{\dot{E}x_u}{\dot{E}x_s}$ Equation 7.	- Dimensions: 1970x970x86 mm. -gross area: 1.91 m^2. -aperture area: 1.73 m^2. -Weight: 41 kg.	Water-glycol.	4.8%
[30]	- Useful exercises in the collector. -Exergy absorbed from solar rays.	$\eta_{ex} = \frac{\dot{E}x_u}{\dot{E}x_s}$ Equation 7.	- Length: 1843 mm. -Width: 1540 mm. -Thickness of glass cover: 10mm. -Absorbent plate thickness: 1.5 mm.	Distilled water.	8%
[31]	- Useful total exergy flow. -Total solar exergy flux.	$\eta_{ex} = \frac{E_{u,t}}{E_{s,t}}$ Equation 7.	- Length: 1,905 m. -Gross area: 2.13 m^2. -Thickness of the plate: 0.002 m. -Pending: 45°.	Water with nanoparticles of Cu.	3.99%
[32]	- Entropy generation rate. -Ambient and solar temperature.	$\eta_{ex} = 1 - \frac{\dot{S}_{gen} T_a}{\left[1 - \left(\frac{T_a}{T_s}\right)\right] \dot{Q}_s}$	- Volume occupied by the collector: 53.2x43.3x8.5 cm^3.	Di Mineralized Water.	6.73%

	- Incident energy in collector area.	Equation 2.	-Absorption area: 0.230 m^2. -Thickness of 2.5 cm.		
[33]	- Flow of exercise in the air. -Exergy flux associated with solar radiation.	$\eta_{ex} = \frac{E_n}{E_s}$ Ecuación 7.	- Length: 1.28 m. -Width: 1 m. -Side and base insulation thickness: 0.5 m.	Air	10.03%
[34]	- Exergy of the desired product. -Exnergy of applied sources.	$\eta_{ex} = 1 - \frac{Ex_d}{Ex_{source}}$ Equation 6.	- Length: 1,955 m. -Width: 1,092 m. -Absorbent plate thickness: 0.18 mm. -Material of plate: copper.	Water.	20.60%
[35]	- Net power of the system. -Exergy input to the collector. -Exergy of solar radiation that reaches the surface of solar systems.	$\psi_{ex} = \frac{\dot{W}_{net}}{\dot{Ex}_{in-eva} + \dot{Ex}_{in-con} + \dot{Ex}_{solar}}$ Equation 12.	- Gross area: 250 m^2. -The collector is assisted by a solar pond to improve its thermal performance.	Water.	11.78%
[36]	- Entropy generation rate. -Ambient and solar temperature. -Incident energy in the collector area.	$\eta_{ex} = 1 - \frac{\dot{S}_{gen} T_a}{\left[1 - \left(\frac{T_a}{T_s}\right)\right] \dot{Q}_s}$ Equation 2.	- Absorption area: 0.375 m^2. -Edge area: 0.1572 m^2. -Glazing thickness: 4 mm. -Weight: 15 kg.	Water with nanoparticles of MgO.	32.23%
[37]	- Entropy generation rate. -Ambient and solar temperature. -Incident energy in the collector area.	$\eta_{ex} = 1 - \frac{\dot{S}_{gen} T_a}{\left[1 - \left(\frac{T_a}{T_s}\right)\right] \dot{Q}_s}$ Equation 2.	- Particle concentration in the range of 0.75 to 1.0% at a mass flow rate of 0.025 to 0.03 kg / s.	Water with nanoparticles of MgO. Water with nanoparticles of CuO.	71.54% 69.11%
[38]	- Exergy output through the fluid. -Surface of the collector. -Normal solar radiation to the collector. -Exergetic factor of solar radiation.	$\eta_{ex} = \frac{\dot{E}x_{out}}{G_t A_c \eta_{max,s}}$ Equation 13.	- Dimension: 250x120x8.6 cm. -Thickness of insulation. 25 mm. -Absorbent plate thickness: 0.5 mm. -Tilt angle: 35°.	Water with nanoparticles of Cu.	10.5%
[39]	- Useful exercises in the collector. -Exergy absorbed from solar rays.	$\eta_{ex} = \frac{\dot{E}x_u}{\dot{E}x_s}$ Equation 7.	- Length: 1,843 m. -Width: 1.54 m. -Height: 0.18 m. -Absorber thickness: 1.5 mm.	Water with nanoparticles of CuO.	11.1%
[40]	- Useful exergy output. -The exergy input to the collector.	$\eta_{ex} = \frac{E_u}{E_{solar}}$ Equation 7.	- Collection area: 1.51 m^2. -Absorbent thickness: 0.006 m. -Insulating thickness: 0.07 m.	Water with nanoparticles of Al_2O_3/Fe.	6.9%

Source: Authors.

In the prepared matrix, the most relevant information of the fifty scientific-technical articles investigated regarding the development of exergy in flat solar collectors is condensed. Mainly, the parameters used in each of the thirteen equations that illustrate the quantification of the exergetic efficiency of the flat plate solar collector used in each article were specified, revealing that 32.5% of the articles investigated are limited to expressing the Efficiency of the exergy as the fraction of the useful exergy in the collector over the exergy absorbed from the solar rays, this being the one most used among the forty models investigated.

Also, you can see the most relevant data for the design of the collector used in each article, this to obtain a general idea of its measurements, for the moment in which you want to take a model as a reference for your own manufacture. Likewise, the working fluids used within the collector of each item are presented, highlighting a greater use of water and water with nanofluids of Al_2O_3, with use in nineteen articles respectively.

Finally, the percentage of exergetic efficiency reached by the flat plate solar collector within each investigated article is exposed; in which higher efficiency ranges are obtained in the collectors that used water, water with nanofluids of Al_2O_3 and water with nanofluids of CuO.

2.3 *Analysis of results*

In the use of solar collectors, it is very important to define the exergetic efficiency, which allows to quantify, in advance, the efficient use of solar energy until reaching equilibrium with a reference environment.

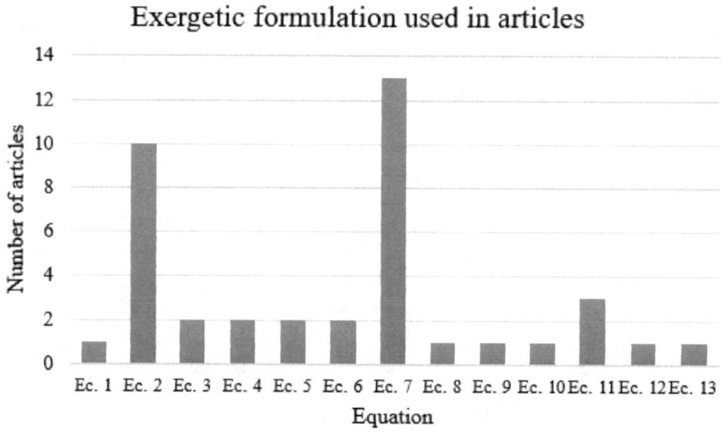

Figure 3. Variation of the number of articles investigated as a function of the exercise efficiency equations in the solar collectors of each investigated model.
Source: Authors.

In Figure 3, you can see the different equations on the exergy efficiency, used in the technical-scientific articles investigated, it is identified that Equation 7 is the most used at a rate of thirteen articles, highlighting that it includes fewer thermodynamic factors and is simpler its solution when applying it in solar collectors; reaching 92% efficiency in an application. Equation 2 was used in ten articles, which although it has more thermodynamic factors in the formula, their development is specified previously and 69% of the useful potential developed with the environment is obtained. Equation 11 has only three adaptations, it is very similar to formula number seven, but it differs in that it uses a difference between the input and output energy, obtaining a maximum efficiency of 24%.

Next, the most prominent and used working fluids in flat plate solar collectors, which were found in the forty articles selected and investigated, are illustrated in a graph.

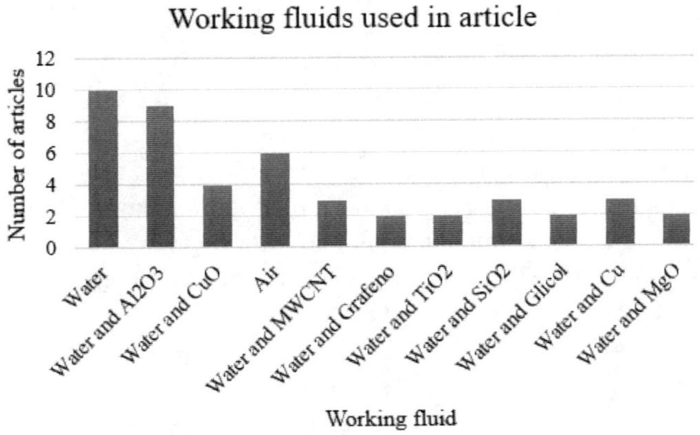

Figure *4*. Variation in the number of articles investigated as a function of the working fluids used in the solar collectors of each investigated model.
Source: Authors.

In Figure 4, the different most used working fluids of the forty articles investigated are observed, it is delimited that 21.74% of the applications use only water as the fluid that receives and transports fluid through the flat solar collector designed in each case. Water with nanofluids of Al_2O_3 was used in 19.57% of the adaptations in the collectors of the different sections, highlighting its use, in order to increase the energetic and exergetic efficiency of the collector designed in the specific article. Only air was used as the working fluid in 13.04% of the sections, with the justification that its obtaining is easy and fast, but not so efficient in terms of exergy. Water with different nanofluids is used in lower percentages of the studies investigated, since they are used in a way that allows evaluating the performance of the collector in terms of several parameters with respect to energy and exercise efficiency.

The working fluids highlighted from the research in the articles are illustrated below, with the variation by ranges of the exergetic efficiency found in the flat plate solar collectors.

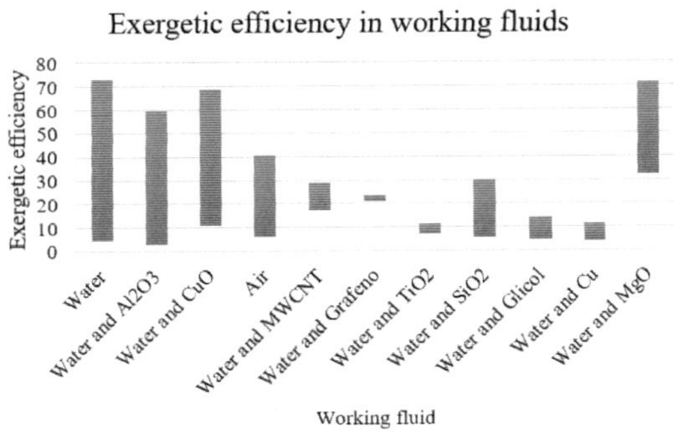

Figure 5. Variation by ranges of the exergetic efficiency as a function of the working fluids used in the solar collectors of each investigated article.
Source: Authors.

In Figure 5, the most commonly used working fluids are illustrated, with the ranges of exergetic efficiency reached in the forty sections investigated. In the first place, it is observed that the widest range is occupied by the use only of water, which varies between 4.62 and 73; Its highest efficiency is due to the fact that its study is carried out by modifying external cycles to the solar collector, such as the Kalina energy cycle and the improvement of the biogas that is integrated in the process. With the use of water with nanofluids of MgO, CuO and Al_2O_3, they obtained a maximum exergy of 71.54%, 69.11% and 59.6% respectively, highlighting that they allow to evaluate and vary the performance of the collector when using them. Air obtained an exergetic efficiency of up to 40%, highlighting that the authors used it as a working fluid due to its easy and fast access. Finally, the combinations of water with other nanofluids had exergies of less than 30%, although these were the least adapted in the different sections, they open a panorama of their application, for subsequent investigations within the same field.

3. Conclusions

The feasibility of carrying out a technical-scientific evaluation of applications in the framework of the development of exergetic efficiency in the capture of solar energy was presented, obtaining as a result, a matrix with a wide panorama of data on thermodynamic parameters, formulation of exergy, most used and prominent designs and working fluids, in flat plate solar collectors, within the last six years.

From the data table, on the investigation in the forty articles destined to the calculation of the exergy in flat solar collectors, it is obtained that the equation of exergetic efficiency most used in the sections, is the one that is limited to expressing the efficiency as the fraction of the exergy useful in the collector over the exergy absorbed from the solar rays, highlighting that it includes fewer thermodynamic factors and its solution is simpler.

In the working fluids applied in the solar collector, it follows that most applications use only water as the fluid that receives and transports fluid through the flat solar collector designed in each case, also, that this same fluid offers a wider range of exergetic efficiency within the solar collector. It is important to note that many authors used the combination of water with nanofluids, in an attempt to evaluate the performance of the collector in terms of various parameters with respect to energy and exergetic efficiency.

Finally, it can be clarified that all the adaptations fully meet their objectives, although some authors managed to obtain an exergy, in flat solar collectors, greater than expected and others lacked very little to fulfill their expected efficiency. From the aforementioned, it can be inferred that it is very important to contribute this research to subsequent projects within the same framework, so that they have a compilation of the experimental background of those carried out in the last six years, and thus, further improve efficiency. obtained in the specific model you are looking for.

References

[1] Y. Tong, H. Lee, W. Kang, and H. Cho, "Energy and exergy comparison of a flat-plate solar collector using water, Al2O3 nanofluid, and CuO nanofluid," *Appl. Therm. Eng.*, vol. 159, no. 113959, 2019.

[2] M. Abuşka and S. Şevik, "Energy, exergy, economic and environmental (4E) analyses of flat-plate and V-groove solar air collectors based on aluminium and copper," *Sol. Energy*, vol. 158, pp. 259–277, 2017.

[3] E. Shojaeizadeh and F. Veysi, "Development of a correlation for parameter controlling using exergy efficiency optimization of an Al2O3/water nanofluid based flat-plate solar collector," *Appl. Therm. Eng.*, vol. 98, pp. 1116–1129, 2016.

[4] M. Eltaweel and A. Abdel-Rehim, "Energy and exergy analysis of a thermosiphon and forced-circulation flat-plate solar collector using MWCNT/Water nanofluid," *Case Stud. Therm. Eng.*, vol. 14, 2019.

[5] S. Verma, A. Tiwari, and D. Chauhan, "Experimental evaluation of flat plate solar collector using nanofluids,"

Energy Convers. Manag., vol. 134, pp. 103–115, 2017.
[6] A. Mete, M. Akif, and A. Turgut, "Thermal performance of a nanofluid-based flat plate solar collector: A transient numerical study," *Appl. Therm. Eng.*, vol. 130, pp. 395–407, 2018.
[7] M. Sharafeldin and G. Grof, "Experimental investigation of flat plate solar collector using CeO2-water nanofluid," *Energy Convers. Manag.*, vol. 155, pp. 32–41, 2018.
[8] M. Sharafeldin, G. Grof, and O. Mahian, "Experimental study on the performance of a flat-plate collector using WO3/Water nanofluids," *Energy*, vol. 141, pp. 2436–2444, 2017.
[9] Z. Said, R. Saidur, M. Sabiha, A. Hepbasli, and N. Rahim, "Energy and exergy efficiency of a flat plate solar collector using pH treated Al2O3 nanofluid," *J. Clean. Prod.*, vol. 112, pp. 3915–3926, 2016.
[10] Z. Said, R. Saidur, and N. Rahim, "Energy and exergy analysis of a flat plate solar collector using different sizes of aluminium oxide based nanofluid," *J. Clean. Prod.*, vol. 133, pp. 518–530, 2016.
[11] M. Ameri and M. Eshaghi, "Exergy and thermal assessment of a Novel system utilizing flat plate collector with the application of nanofluid in porous media at a constant magnetic field," *Therm. Sci. Eng. Prog.*, vol. 8, pp. 223–235, 2018.
[12] A. Alklaibi, L. Sundar, and A. Sousa, "Experimental analysis of exergy efficiency and entropy generation of diamond/water nanofluids flow in a thermosyphon flat plate solar collector," *Int. Commun. Heat Mass Transf.*, vol. 120, 2021.
[13] L. Kumar, S. Kazi, H. Masjuki, M. Zubir, A. Jahan, and C. Bhinitha, "Energy, exergy and economic analysis of liquid flat-plate solar collector using green covalent functionalized graphene nanoplatelets," *Appl. Therm. Eng.*, vol. 192, 2021.
[14] T. Ural, "Experimental performance assessment of a new flat-plate solar air collector having textile fabric as absorber using energy and exergy analyses," *Energy*, vol. 188, 2019.
[15] S. Khanmohammadi, M. Baseri, P. Ahmadi, A. Al-Rashed, and M. Afrand, "Proposal of a novel integrated ocean thermal energy conversion system with flat plate solar collectors and thermoelectric generators: Energy, exergy and environmental analyses," *J. Clean. Prod.*, vol. 256, 2020.
[16] K. Balaji, S. Iniyan, and M. Swami, "Exergy, economic and environmental analysis of forced circulation flat plate solar collector using heat transfer enhancer in riser tube," *J. Clean. Prod.*, vol. 171, pp. 1118–1127, 2018.
[17] A. Allouhi, M. Amine, M. Buker, T. Kousksou, and A. Jamail, "Forced-circulation solar water heating system using heat pipe-flat plate collectors: Energy and exergy analysis," *Energy*, vol. 180, pp. 429–443, 2019.
[18] A. Mortazavi and M. Ameri, "Conventional and advanced exergy analysis of solar flat plate air collector," *Energy*, vol. 142, pp. 277–288, 2018.
[19] M. Ammar, A. Mokni, H. Mhiri, and P. Bournot, "Performance optimization of flat plate solar collector through the integration of different slats arrangements made of transparent insulation material," *Sustain. Energy Technol. Assessments*, vol. 46, 2021.
[20] M. Mehrpooya, B. Ghorbani, and S. Hosseini, "Developing and exergetic performance assessment of biogas upgrading process driven by flat plate solar collectors coupled with Kalina power cycle," *Energy Convers. Manag.*, vol. 181, pp. 398–413, 2019.
[21] M. Edalatpour and J. Solano, "Thermal-hydraulic characteristics and exergy performance in tube-on-sheet flat plate solar collectors: Effects of nanofluids and mixed convection," *Int. J. Therm. Sci.*, vol. 118, pp. 397–409, 2017.
[22] M. Ashouri, M. H. Ahmadi, M. Feidt, F. Astaraei, and M. Razi, "Exergy and energy analysis of a regenerative organic Rankine cycle based on flat plate solar collectors," *Mech. Ind.*, vol. 18, 2017.
[23] G. Datong *et al.*, "Experimental and numerical analysis of an efficiently optimized evacuated flat plate solar collector under medium temperature," *Appl. Energy*, vol. 269, 2020.
[24] P. Ouyang, Y.-P. Xu, L.-Y. Qi, S.-M. Xing, and H. Fooladi, "Comprehensive evaluation of flat plate and parabolic dish solar collectors' performance using different operating fluids and MWCNT nanofluid in different climatic conditions," *Energy Reports*, vol. 7, pp. 2436–2451, 2021.
[25] R. Loni and S. Gorjian, "Thermodynamic analysis of a solar dish receiver using different nanofluids," *Energy*, vol. 133, pp. 749–760, 2017.
[26] V. Stefanovic and C. Tzivanidis, "A detailed parametric analysis of a solar dish collector," *Sustain. Energy Technol. Assessments*, vol. 25, pp. 99–110, 2018.
[27] N. Akram and W. Sarsam, "Experimental investigations of the performance of a flat-plate solar collector using carbon and metal oxides based nanofluids," *Energy*, vol. 227, 2021.

[28] M. Hanif, K. Khan, M. Khan, M. Ramzan, and Abdurab, "Energy, exergy and efficiency analysis of a flat plate solar collector used as air heater," *Sains Malaysiana*, vol. 47, no. 5, pp. 1061–1067, 2018.
[29] A. Gheorghian, T. Prisecaru, and M. Ciobanu, "Energy and exergy analysis of a flat plate collector using collector test equations," *E3S Web Conf.*, vol. 180, 2020.
[30] A. Allouhi and M. Amine, "Effect analysis on energetic, exergetic and financial performance of a flat plate collector with heat pipes," *Energy Convers. Manag.*, vol. 195, pp. 274–289, 2019.
[31] E. Bellos and C. Tzivanidis, "Performance analysis and optimization of an absorption chiller driven by nanofluid based solar flat plate collector," *J. Clean. Prod.*, vol. 174, pp. 256–272, 2018.
[32] S. Verma and N. Upadhyay, "'Performance comparison of innovative spiral shaped solar collector design with conventional flat plate solar collector,'" *Energy*, vol. 194, 2020.
[33] A. Kumar and A. Layek, "Energetic and exergetic performance evaluation of solar air heater with twisted rib roughness on absorber plate," *J. Clean. Prod.*, vol. 232, pp. 617–628, 2019.
[34] M. Ameri and M. Eshaghi, "A novel configuration of reverse osmosis, humidification–dehumidification and flat plate collector: Modeling and exergy analysis," *Appl. Therm. Eng.*, vol. 103, pp. 855–873, 2016.
[35] M. Erden and I. Dincer, "Performance investigation of hydrogen production by the flat-plate collectors assisted by a solar pond," *Int. J. Hydrogen Energy*, vol. 42, pp. 2522–2529, 2017.
[36] S. Verma and D. Chauhan, "Performance augmentation in flat plate solar collector using MgO/water nanofluid," *Energy Convers. Manag.*, vol. 124, pp. 607–617, 2016.
[37] S. Verma and D. Chauhan, "Performance analysis of hybrid nanofluids in flat plate solar collector as an advanced working fluid," *Sol. Energy*, vol. 167, pp. 231–241, 2018.
[38] S. Shamshirgaran and H. Al-Kayiem, "Upper limits for the work extraction by nanofluid-filled selective flat-plate solar collectors," *Energy*, vol. 160, pp. 875–885, 2018.
[39] A. Allouhi and M. Amine, "Heat pipe flat plate solar collectors operating with nanofluids," *Sol. Energy Mater. Sol. Cells*, vol. 219, 2021.
[40] E. Okonkwo and T. Al-Ansari, "Thermodynamic evaluation and optimization of a flat plate collector operating with alumina and iron mono and hybrid nanofluids," *Sustain. Energy Technol. Assessments*, vol. 37, 2020.

Performance analysis of PKM 60 peripheral pumps as a consequence of impeller wearing.

M A Duran-Sarmiento[1,2], L F García-Rodríguez[1], C L Sandoval-Rodríguez[1], J G Ascanio-Villabona[1,2] and O Lengerke-Pérez[1]

[1] Design and materials research group (DIMAT), Automation and Control Energy Systems Research Group (GISEAC), Electromechanical engineering, Faculty of Natural Sciences and Engineering, Technological Units of Santander (UTS), Calle de los estudiantes No 9-82, 680005, Bucaramanga, Colombia.
[2] Energy in Building research group (ENEDI), Doctoral Program in Energy Efficiency and Sustainability in Engineering and Architecture, Department of Machines and Thermal Engines, University of the Basque Country, UPV/EHU, Engineer Torres Quevedo Plaza, 1, 48013, Bilbao, Spain.

aduran@correo.uts.edu.co

Abstract. The efficient transformation of energy in pumping systems plays a fundamental role in the industry, these pumping systems represent a significant energy consumption, if the pumps present wear in their impeller, they will significantly decrease their capacity to communicate energy to the fluid, that in other words, the system will require more energy to compensate for this decrease in efficiency, causing an increase in operating costs. This work presents a comparison of the performance of two identical pumps with different impellers, a new impeller and an impeller subjected to wear, which has been emulated by machining the impeller in its radial component. The results show a decrease in the peak net height and power communicated to the fluid of 55% and 63% respectively. The percentages of decrease in the capacity to transfer energy to the fluid obtained in the test are not equivalent to the percentage of diameter reduction, which presented a decrease from 56.5 mm to 59.6 mm, for a percentage of diameter decrease equivalent to 5.2%.

Keywords: efficiency, pump, hydraulic systems, curve comparison, power decresase

1. Introduction.

Pumping systems satisfy industrial needs at different levels, transforming mechanical energy into useful energy for the fluid, this energy is used to defeat friction in the pipeline and fittings to provide pressure and flow according to the need. As a consequence of wear and tear, pumps lose the ability to lift the same flow and pressure, this phenomenon (sometimes ignored) can significantly increase energy consumption in the system [1].

For this reason, impeller wear must be considered in order to determine acceptable operating margins [2], This wear can occur for several reasons, such as friction with water, corrosion and cavitation [3] and others; in this case, only radial wear is analyzed [4].

Pumps constitute 22% of the energy consumption in fluid transport [5], For this reason, it is essential to study the effect of wear in pumps, as such wear affects the capacity to transfer energy to the fluid, taking into consider that energy consumption worldwide is expected to increase by 30% from 2016 to 2030.

This paper presents the results of the comparison of the characteristic curves of useful head and power communicated to the fluid, as performance curves [6] of the pump with an intact impeller compared to a pump with a radially worn impeller emulated by machining through the implementation of the PKM60 impeller performance test module, the fluid power delivered as a function of flow rate for each of the impellers is also estimated.

2. Work development

In order to estimate the change in the pressure vs. flow curve of the PKM 60 pump, it is necessary to simulate the wear of the impeller using mechanized techniques to compare its operation with the impeller in optimal conditions. This comparison was carried out in a module designed to perform the tests on the pump.

2.1. Design of test module

It was established that the best way to compare the performance of the wear-induced impeller and the non-wear impeller is to implement a system that can handle separate pumps, one for each impeller [7].

The module has instrumentation that can measure the suction and discharge pressure of each pump and a rotameter to measure the flow rate.

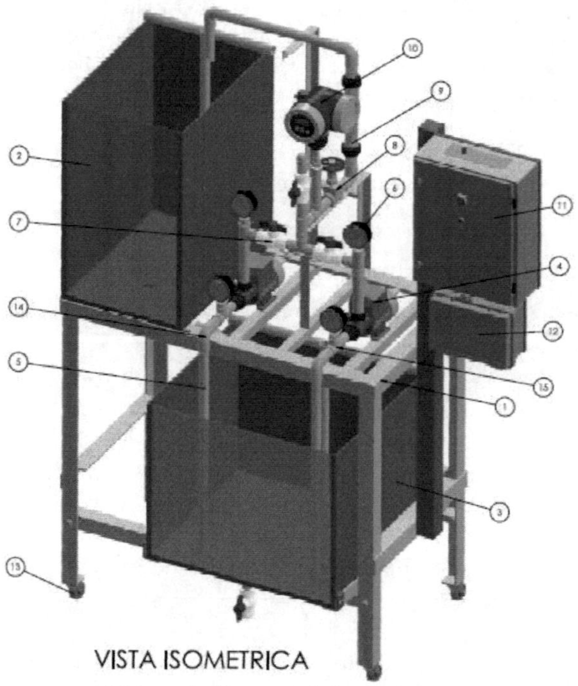

Figure 1. Test module design

IOP Conf. Series: Materials Science and Engineering 1253 (2022) 012010 doi:10.1088/1757-899X/1253/1/012010

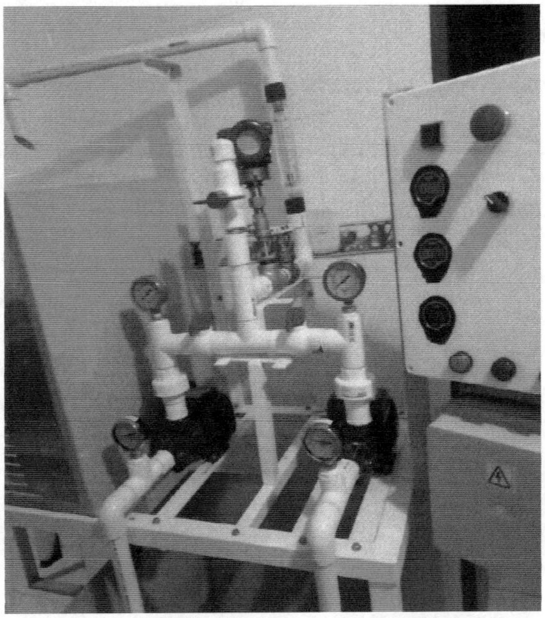

Figure 2. Impeller wear test module

The impellers with and without wear used in the study are 56.5 mm and 59.6 mm respectively, emulating a radial wear of 3.1 mm.

New Impeller **Impaired Impeller**

Figure 3. Pump PKM 60 impellers

The module has a suction and discharge tank, the discharge tank is calibrated to compare the rotameter measurement with the actual discharge flow value.

2.2. Testing methodology

The correct power supply to the module and the protection of mechanical, electrical and electronic equipment must be ensured, as well as the priming of the pumps [8]. The test was carried out by starting the pump with the impeller new or wear-free, taking into account that excessive wear is negative for the pump performance [9], the variables measured in this test were flow rate, pump suction and discharge pressure, as well as current and voltage.

The flow rate was modified by a throttling valve at the pump discharge to take pressure data for a flow rate range from 5 lpm to 30 lpm, and these values and their corresponding suction and discharge pressures, current and voltage were registered. After performing the test with the pump with the new or wear-free impeller, the same procedure was performed with the pump having the impeller with the emulated wear. With the data of the pressures at the pump inlet and outlet, the net head can be determined by equation 1 and the hydraulic power by equation 2 [10].

$$H = \frac{p_2 - p_1}{\rho g} \tag{1}$$

$$p_w = \rho g Q(\tag{2}$$

Where p_2 is the pressure at the discharge, p_1 is the pressure at the suction, ρ is the density of the test fluid, g is the acceleration of gravity and Q is the pump flow rate.

3. Results and discussions.

The tests were done for each of the pumps, taking into consideration the variation of the flow rate between 5 lpm and 30 lpm, the pressures at the suction and discharge of the pump, as well as the current and voltage of the pump. The results obtained in the hydrodynamic tests on the pump with the new impeller and the pump with the wear induced impeller were recorded in Table 1 and 2 respectively.

Table 1. New impeller data

New Impeller				
Q(lpm)	**Inlet pressure (psi)**	**Outlet pressure(psi)**	**Voltage (V)**	**Current (A)**
11,356	-4	43	125	5,4
15	-4,32	37	126	5,2
18,927	-4,68	31	125	5
20	-4,92	29	125	4,9
22,713	-4,97	25	125	4,8
25	-4,99	21	125	4,7
26,498	-5	18	126	4,6
30	-5,12	13	126	4,5
34,06	-5,23	3	126	4,4

Source: Author

Table 2. Impaired impeller data

Impaired Impeller				
Q(lpm)	**Inlet pressure (psi)**	**Outlet pressure(psi)**	**Voltage (V)**	**Current (A)**
0	0	26	124	4,7
7,5708	-3,85	20	124	4,6
10	-3,99	18	124	4,5
11,356	-4,58	16,5	124	4,5
15	-4,67	13	124	4,5
18,92	-4,89	10	124	4,4
20	-4,95	8,5	124	4,5
22,71	-5,03	6	124	4,4
25	-5,11	4,8	124	4,4
26,49	-5,53	3,8	124	4,4
28,39	-6	2,4	124	4,5

Source. Author

The values obtained during the tests were used to estimate the net head values using equation 1 and represented in a graph, the heads corresponding to the same flow rates in both tests were selected. It is observed in the diagram the significant decrease in the net head, it should be considered that the decrease of the diameter from 59.6 mm to 56.5 mm, represents an affectation of 5.2 % of the impeller diameter, but it does not represent the same percentage of decrease in the useful area of the impeller, the maximum head was reduced by approximately 55.13 %.

IOP Conf. Series: Materials Science and Engineering 1253 (2022) 012010 doi:10.1088/1757-899X/1253/1/012010

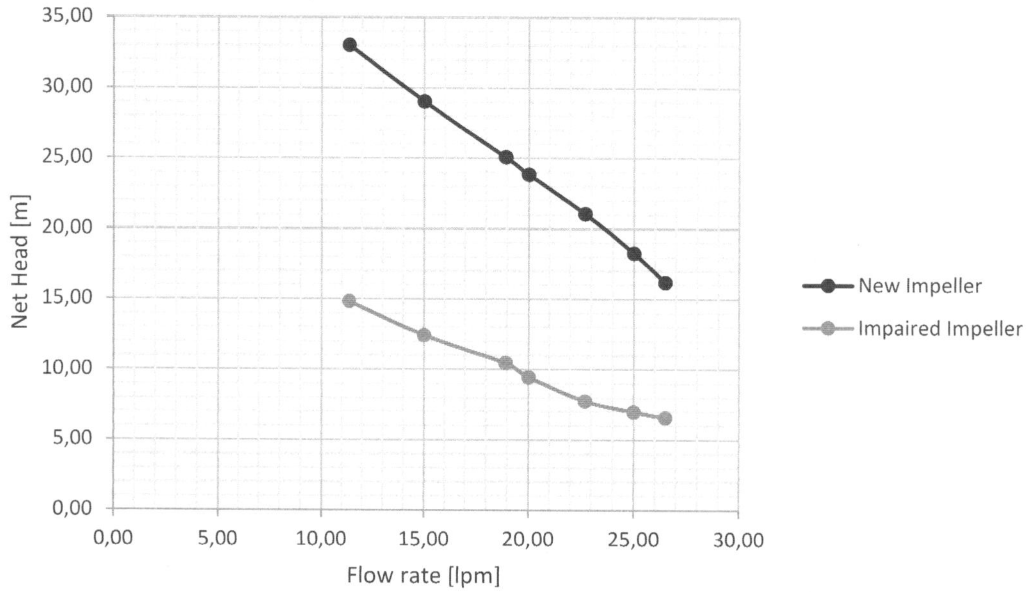

Figure 4. Net head comparison

An analysis of the power communicated to the fluid by each of the pumps was also performed

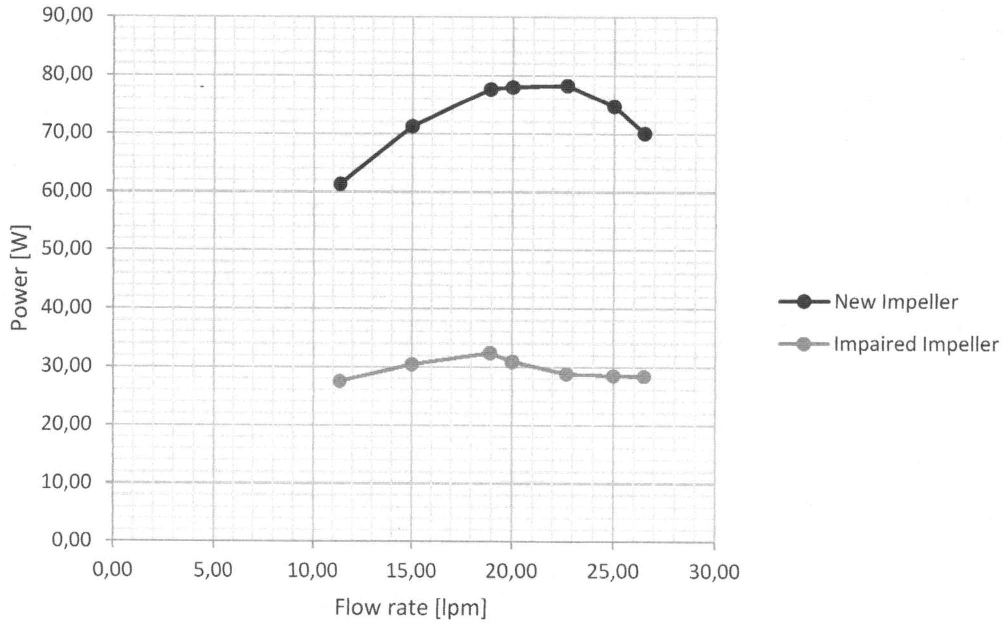

Figure 5. Net power communicated to fluid

The tendency in the behavior of the power communicated to the fluid shows a significant decrease, the peak power communicated was reduced by approximately 63.19%.

It should be noted that the decrease in net head and power communicated to the fluid show higher reduction percentages than the percentage reduction in the area of contact with the fluid, taking into consider that both energy, power and head depend on the area, where the radial component of the absolute velocity developed in the impeller depends on the area, showing that there is no linear relationship between these variables.

$$c_{2m} = \frac{Q}{2 * \pi * r_2 * b_2} \tag{3}$$

The radial component of the absolute velocity is related to the height, power and communicated energy, as the radial component(c_{2m}) increases, the peripheral component of the absolute velocity (c_{2u}) decreases and it is the peripheral velocity that significantly conditions the magnitude of all the energy communicated to the fluid (see equation 4 and 5).

$$Y_u = C_{2u} * U_2 - C_{1u} * U_1 \tag{4}$$

$$Y_u = H_u * g \tag{5}$$

4. Conclusions and recommendations.

It is reflected the impact of wear in the unemployment of pumping equipment, this wear can occur for several reasons, friction with water, corrosion, cavitation among others, the emulation of wear for this study, was only in the radial component, it is suggested for further studies the analysis of wear in the axial component.

Pump performance is significantly affected by impeller wear; peak net head and power delivered to the fluid decreased 55% and 63%, respectively.

The percentage reduction in peak net head (55%) and power delivered to the fluid (63%) do not equal the percentage reduction in impeller diameter (5.2%) due to wear.

It is recommended for further studies to test with different types of impellers for a more comprehensive comparison.

5. References.

[1] X. Li, B. Chen, X. Luo, and Z. Zhu, "Effects of flow pattern on hydraulic performance and energy conversion characterisation in a centrifugal pump," *Renew. Energy*, vol. 151, pp. 475–487, 2020, doi: 10.1016/j.renene.2019.11.049.

[2] T. Lin, Z. Zhu, X. Li, J. Li, and Y. Lin, "Theoretical, experimental, and numerical methods to predict the best efficiency point of centrifugal pump as turbine," *Renew. Energy*, vol. 168, pp. 31–44, 2021, doi: 10.1016/j.renene.2020.12.040.

[3] O. Glovatsky, R. Ergashev, A. Saparov, M. Berdiev, and B. Shodiev, "Cavitation-abrasive wear working collectors of pumps," *IOP Conf. Ser. Mater. Sci. Eng.*, vol. 869, no. 4, 2020, doi: 10.1088/1757-899X/869/4/042006.

[4] M. DaqiqShirazi, R. Torabi, A. Riasi, and S. A. Nourbakhsh, "The effect of wear ring clearance on flow field in the impeller sidewall gap and efficiency of a low specific speed centrifugal pump," *Proc. Inst. Mech. Eng. Part C J. Mech. Eng. Sci.*, vol. 232, no. 17, pp. 3062–3073,

2018, doi: 10.1177/0954406217729420.

[5] V. K. Arun Shankar, S. Umashankar, S. Paramasivam, and N. Hanigovszki, "A comprehensive review on energy efficiency enhancement initiatives in centrifugal pumping system," *Appl. Energy*, vol. 181, pp. 495–513, 2016, doi: 10.1016/j.apenergy.2016.08.070.

[6] X. Li, H. Chen, B. Chen, X. Luo, B. Yang, and Z. Zhu, "Investigation of flow pattern and hydraulic performance of a centrifugal pump impeller through the PIV method," *Renew. Energy*, vol. 162, pp. 561–574, 2020, doi: 10.1016/j.renene.2020.08.103.

[7] N. Zhang, X. Liu, B. Gao, X. Wang, and B. Xia, "Effects of modifying the blade trailing edge profile on unsteady pressure pulsations and flow structures in a centrifugal pump," *Int. J. Heat Fluid Flow*, vol. 75, no. April 2018, pp. 227–238, 2019, doi: 10.1016/j.ijheatfluidflow.2019.01.009.

[8] D. Wu, Z. Zhu, Y. Ren, Y. Gu, and P. Zhou, "Influence of blade profile on energy loss of sewage self-priming pump," *J. Brazilian Soc. Mech. Sci. Eng.*, vol. 41, no. 10, 2019, doi: 10.1007/s40430-019-1978-2.

[9] G. Peng, F. Fan, L. Zhou, X. Huang, and J. Ma, "Optimal hydraulic design to minimize erosive wear in a centrifugal slurry pump impeller," *Eng. Fail. Anal.*, vol. 120, no. November, p. 105105, 2021, doi: 10.1016/j.engfailanal.2020.105105.

[10] Y. L. Zhang, Z. C. Zhu, H. S. Dou, and B. L. Cui, "A generalized euler equation to predict theoretical head of turbomachinery," *Int. J. Fluid Mech. Res.*, vol. 42, no. 1, pp. 26–38, 2015, doi: 10.1615/InterJFluidMechRes.v42.i1.30.

IOP Conf. Series: Materials Science and Engineering 1253 (2022) 012011 doi:10.1088/1757-899X/1253/1/012011

Development of a vehicle to automatically load, transport and store the product of a blackberry crop

C.A. Angulo- Julio[1*], C.L. Sandoval-Rodríguez[1|-2], O. Lengerke-Perez[1], B.E Tarazona-Romero y J.G. Ascanio-Villabona[1-2].

[1]**Faculty of Natural Sciences and Engineering, Unidades Tecnológicas de Santander, Bucaramanga, Colombia**

[2]**University of the Basque Country, Bilbao, Spain**

carlosaj@correo.uts.edu.co

Abstract. Harvesting the blackberry crop is a delicate process because it must be carried out in the first hours of the day in shallow containers and without overhandling to prevent deterioration of the fruit and its fermentation. In this work a prototype of an automated vehicle is developed to improve the harvesting process of the crop reducing the time without affecting the quality of the blackberry. CAD tools were used to design the structure of the prototype and Arduino boards were used for its control, while communication between the vehicle and the monitoring station was established with Xbee devices. The vehicle moves carrying the harvest product in baskets between the collection points and the warehouse. Automated loading and unloading systems eliminate double handling of the products and require less human intervention. Hence, the implementation of the vehicle for loading, transporting, and unloading the product of the blackberry harvest generates savings in production costs and improves collection times since workers will be able to focus on the selection, harvesting and classification process.

1. Introduction

The blackberry is a fruit very susceptible to bruising, so special care must be taken in the harvesting process [1], [2]. The reap of the fruits from the plant must be done with gentle movements and then place them in a container without hitting the fruits and without accumulating several in the hands to avoid the growth of fungi [3]. It is recommended that the basket have separations to classify the blackberries, although it is also possible to collect all the fruits in the same container and then do the classification.

To preserve the quality of the product to the maximum, the harvest should be carried out in the first hours of the day, taking advantage of the morning dew because once it disappears it is easier for the fruit to ferment [4]. Harvesting must be done carefully selecting the fruits, avoiding picking those that are still green. It is also important that they are not handled excessively or accumulate on the hands [5].

To avoid bruising the fruits due to the excess weight generated by their stacking, the container in which they are collected should not be too deep [2]. This implies that growers must constantly move from the pickup site to the warehouse site, reducing the effective time in which they can collect fruit from the plants [6]. By developing an automated vehicle for loading, transporting, and unloading the product collected by growers, it is intended to reduce the time and costs of the harvesting process [7]. This considering that the workers will be able to concentrate only on the selection and classification of the fruits without taking care of their transport to the warehouse.

2. Characteristics of blackberry crops

The blackberry (*rubus glaucus benth*) is a fruit belonging to the rosaceae family that is characterized by being shrubby with creeping or semi-erect stems. Its fruit is an aggregate of drupes, each with seeds inside [8]. The plant presents permanent flowering and fruiting and is cultivated in cold regions with temperatures of 16 ° C to 18 ° C and altitudes that range between 1,200 m and 3,500 m [9], [10].

Blackberry crops have great agroindustrial relevance in Colombia since the fruit is a raw material for the preparation of jams and concentrate juices, among others [11]. They are spread over sixteen Departments, Santander, Cundinamarca and Antioquia being the most representative with 57.93% of national production [12]. In the last years, it has shown a growing trend both at the level of production and the area harvested in response to the greater demand from the national agribusiness [13].

In Colombia, the yield per hectare under Colombian production conditions varies from six to sixteen tons, with a national average of 11 tons per hectare. However, in highly technical crops, higher yields have been reported. According to Grijalba [14], a well-managed crop can produce from 18 to 20 tons / ha per year, January, February, July, August and December being the months of greatest harvesting. On the other hand, according to Ramirez [15], if 2,500 plants per hectare are established from 18 months onwards and depending on the care given to the crop, productions of 14 to 16 tons per hectare can be achieved in a productive year.

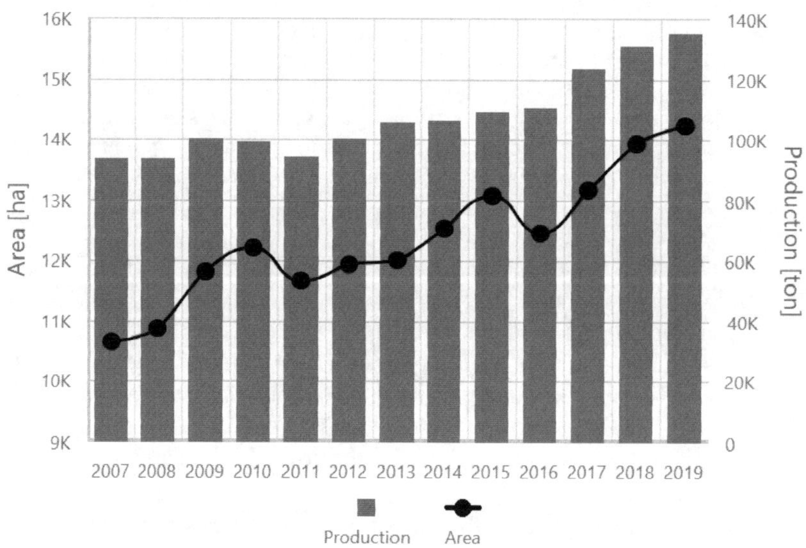

Figure 1. Blackberry production and harvested area in Colombia [13]

3. Prototype development

Alternative industrial systems, in a simple and accessible way, allow a wide field of measurement procedures that can be used in robust regimes compared to other controls [16]. The operation of fruit harvesting machines is based on the method of shaking the branches of the plant until the fruit falls on a platform. In these cases, the system must be designed in such a way that it allows it to go through the cultivation in production lines to guarantee efficiency in the harvest while minimizing the loss of

IOP Conf. Series: Materials Science and Engineering 1253 (2022) 012011 doi:10.1088/1757-899X/1253/1/012011

fruits [17]. When using this system, the fruits could be mistreated when they are plucked or when they fall. Therefore, given the delicacy of the blackberry, harvesting through this type of system would significantly affect the quality of the product. Thus, it was concluded that it was not feasible to use this type of harvesting system in blackberry crops.

Instead, here we develop a vehicle that transports to the storage warehouse the fruits manually picked by the collectors. In this way, workers will be able to concentrate solely on the collection tasks, leaving aside the transport task. To make the harvesting process more efficient and productive, we designed an automated vehicle prototype.

The scaled vehicle has a route system built considering each collection station and has characteristics to transport the baskets with the collected blackberry. In this way, it is intended to improve the transport, loading, and unloading of baskets in blackberry crops to save costs and time. The solution developed and the aspects considered in the definition of the necessary components for the development of the prototype are presented below.

3.1. Structural design

For the construction of the prototype, a model was previously designed to represent the crop, for which the spaces occupied for the harvest stations and the storage site were taken into account, as well as the spaces that will be allocated for the mobilization of the vehicle. collector. Figure 2 shows the model used to represent the crop of blackberries. The green zones in the figure represent the five harvest stations considered in our system.

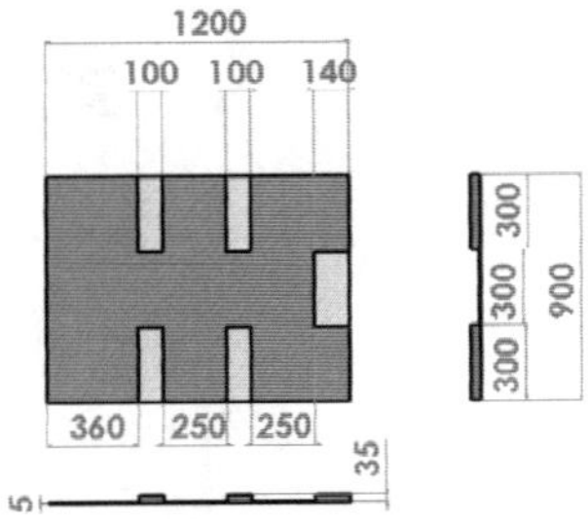

Figure 2. Representation of a blackberry crop (dimensions in cm).

The design of the prototype was carried out using SolidWorks software considering structure, outlines, and dimensions. The external structure of the collecting vehicle is shown in Figure 3, its dimensions are displayed in Figure 4, while the design of the parts that make up the prototype with the respective dimensions and auxiliary views is presented in Figure 5.

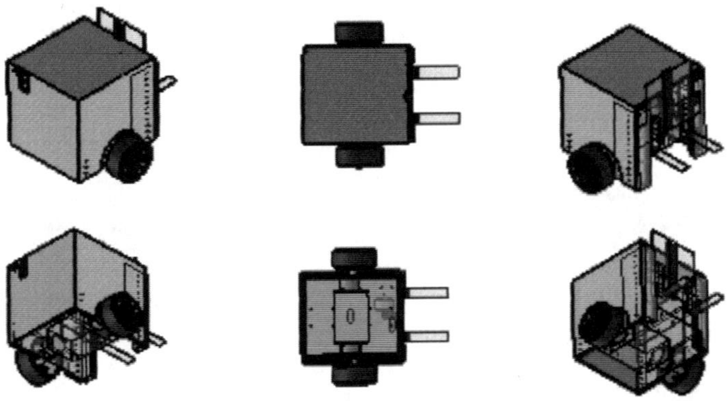

Figure 3. External structure of the vehicle.

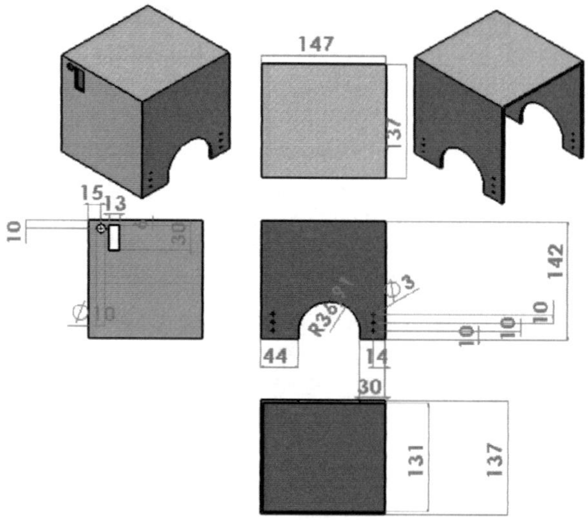

Figure 4. External dimension of the vehicle.

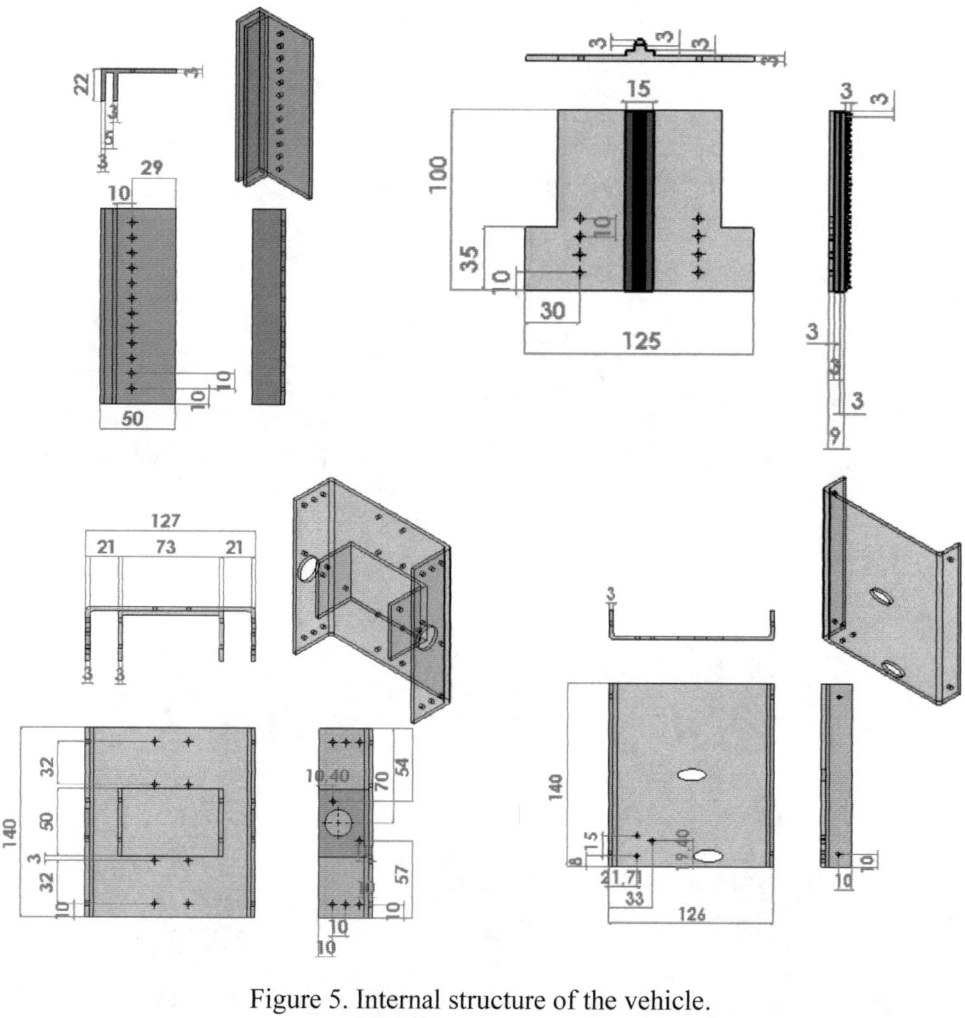

Figure 5. Internal structure of the vehicle.

4. Internal components

Controlling the trajectory and location of the vehicle is a very important factor in the design of the prototype. A control that is almost insensitive to disturbances can be obtained using state-of-the-art software and with the help of mathematical calculations [18]. Among the possibilities available to determine the position of the vehicle is the location control by means of global positioning systems (GPS) - with proximity sensors and creating a trajectory guide - or by controlling the location remotely through wireless communication.

Although the use of GPS may be feasible in the real case, for the implementation of the prototype it was discarded since it did not meet expectations due to the dimensions of the model and the margin of error in the position offered by the GPS.

The general behavior of the system is given by the GRAFCET shown in the figure. The operation consists of five macro steps to load the baskets with the fruit to the vehicle (one for each recollection station) and an unloading step at the storage point. The vehicle will go to one of the collection stations according to the request made by the collectors. Once there, it will load the baskets with the blackberries and then it will move to the storage point, where the baskets will be unloaded.

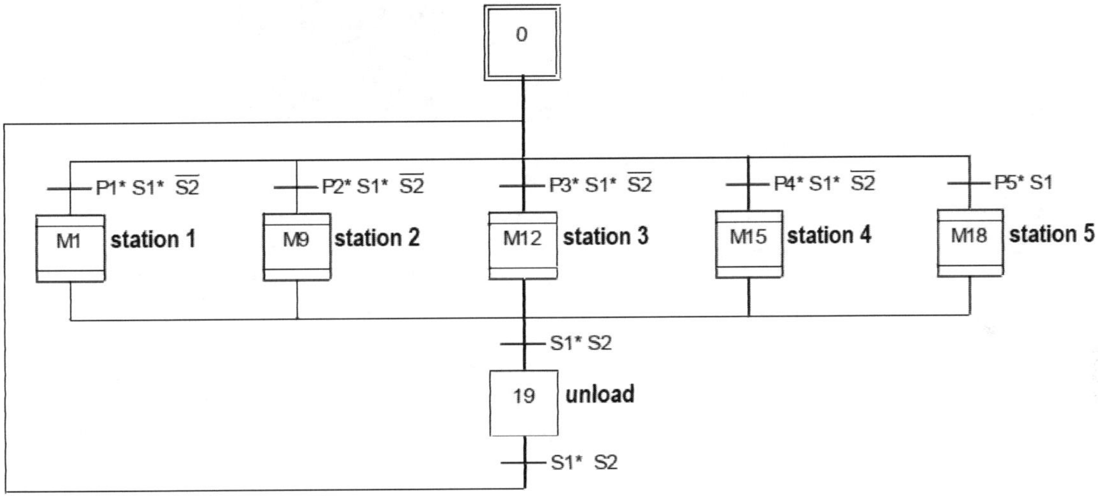

Figure 6. General behavior of the system.

The system has Xbee modules, Arduino boards and sensors in order to control the vehicle and establish its location in real time.

4.1. Xbee technology

There are different technologies to establish communication between the vehicle and the control station. Bluetooth technology offers a connection range of up to 10 meters in optimal conditions, while with WiFi technology greater ranges can be achieved as long as there is a network coverage [19]. Therefore, to establish communication between the vehicle and the control center, Xbee S2 modules was chosen. With Xbee technology, modules can be used to cover large communication distances, and there are models with which distances greater than 1 km can be reached. If necessary, Xbee technology allows the extension of coverage by creating a network interconnecting more modules. Xbee boards can be configured to establish bidirectional communication and there are modules compatible with Arduino boards and Labview software. [19], [20]

4.2. Arduino boards

The implemented system has Arduino Mega boards to control the motors of the loading / unloading system of the baskets with blackberries, activate the motors to move the vehicle, as well as establish the status of the process, that is, to manage requests from the collectors in each of the stations, to know the status of the vehicle, to monitor of the established route and synchronization with the command center. Figure 7 shows the connection scheme between an Xbee S2 module and an Arduino Mega board.

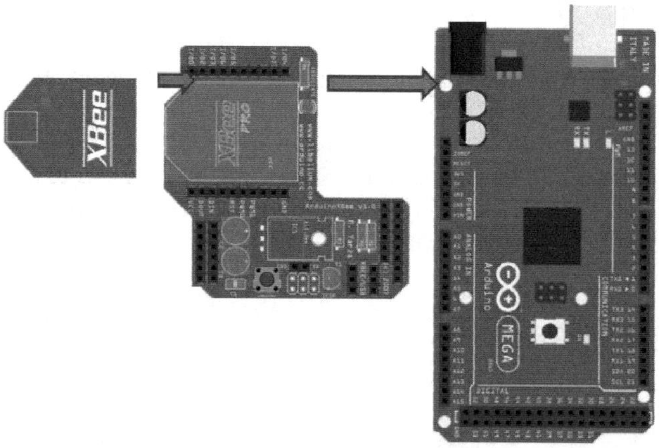

Figure 7. Xbee S2 and Arduino Mega boards setup.

4.3. Command center

For the system command center, we developed an application so as to real-time monitoring the general status of the harvesting process. LabVIEW offers a graphical programming environment to create an interactive interface of the whole process and to read signals and act in real time [21]. Figure 8 shows the front panel of the application made in Labview. There it can be seen the different collection stations, the unloading sites at the storage point and indicators that show the location of the vehicle, if any collector made a request for the vehicle to collect the harvested fruit and if the availability of the sites arranged for the unloading of the baskets with the blackberry. The application communicates with the vehicle and the different elements that make up the system to determine the route of the vehicle and the actions of the process.

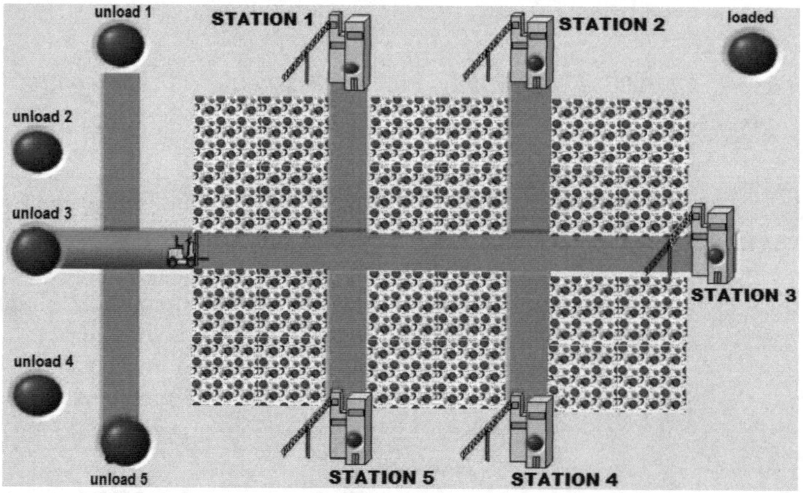

Figure 8. Front panel of command center application.

5. Effectiveness of the prototype

We carried out some tests to analyze the operation of the vehicle within the blackberry harvest process. The Table 1 shows the results of the tests carried out to determine the effect caused by the weight of the blackberries in terms of the time it takes for the vehicle to carry out the necessary actions to move over the crop. Likewise, the table summarizes the results of the tests carried out to measure the misalignment distance of the vehicle in relation to the desired position whether it was loaded or not.

Those results show that the vehicle presents a minimum misalignment in the routes that it covers towards each of the stations. This is because the tires do not have an optimal balance and becomes more evident due to the weight added when the baskets are loaded.

Table 1. Time in seconds to move the vehicle over each station whether it was loaded or unloaded.

Station	Vehicle	Go forward	Go backward	Turn right	Turn left	TOTAL
1	loaded	21	21	22	0	**64**
	unloaded	7	7	22	0	**36**
2	loaded	35	35	22	0	**92**
	unloaded	7	7	22	22	**73**
3	loaded	35	35	0	0	**70**
	unloaded	10.5	10.5	22	22	**65**
4	loaded	35	35	0	22	**92**
	unloaded	7	7	22	22	**73**
5	loaded	21	21	0	22	**64**
	unloaded	7	7	0	22	**36**

Table 2. Misalignment distance (in cm).

	Vehicle loaded	Vehicle unloaded
Station 1	3.25	7.95
Station 2	4.35	8.50
Station 3	3.30	9.50
Station 4	4.15	8.65
Station 5	3.60	7.30

6. Conclusions

Automation is essential for processes to be more efficient. This efficiency is what is sought to improve in blackberry crops with the implementation of the system that allows the loading, unloading and transport of blackberry. By implementing this system, improvements are generated in the production process in terms of quality and quantity of blackberry because the workers will spend more time choosing and collecting the fruits with the best characteristics of the plants, while the automated vehicle pick up the baskets full of blackberries and take them from the stations to the warehouse. Additionally, savings are generated in production costs proportional to the number of cultivated hectares and harvest times are improved thanks to the fact that workers will not spend time taking the harvest to the warehouse, but instead will leave the baskets with blackberry at the collection points strategically located throughout the crop.

The application developed for the operation center allows monitoring and control of the general blackberry harvesting process, being able to remotely manage the vehicle by establishing its movements. Likewise, the actions that the vehicle is carrying out automatically can be traced in real time.

The automated vehicle can not only be used on blackberry crops but also on other types of crops. It could also be useful in processes where it is necessary to transport heavy items that exceed the load limit per person, such as supermarkets, warehouses, or factories. However, the possible effects that the weight of the load may entail must be considered in order to adjust the control actions.

References

[1] J. B. Samtani and M. M. Kushad, "A Longer Marketing Life for Blackberry and Raspberry Fruit," *Virginia Cooperative Extension*, no. 423–701, May 2015, [Online]. Available: https://vtechworks.lib.vt.edu/handle/10919/75450

[2] S. Horvitz, *Postharvest Handling of Berries*. IntechOpen, 2017. doi: 10.5772/intechopen.69073.

[3] J. D. Ramírez Q, I. D. Aristizábal T, and J. I. Restrepo F, "Blackberry conservation through the application of edible coating of aloe vera mucilaginous gel," *Vitae*, vol. 20, no. 3, pp. 172–183, Dec. 2013.

[4] J. McCoy, J. Clarke, A. Salgado, and A. Jecmen, "Evaluation of harvest time/temperature and storage temperature on postharvest incidence of red drupelet reversion development and firmness of blackberry (Rubus L. subgenus Rubus Watson)," *Discovery, The Student Journal of Dale Bumpers College of Agricultural, Food and Life Sciences*, vol. 17, no. 1, pp. 59–65, Jan. 2016.

[5] B. Lawrence and J. C. Melgar, "Harvest, Handling, and Storage Recommendations for Improving Postharvest Quality of Blackberry Cultivars," *HortTechnology*, vol. 28, no. 5, pp. 578–583, Oct. 2018, doi: 10.21273/HORTTECH04062-18.

[6] Feucht-Obsttechnik, "Fruit harvesting technology," *Feucht Obsttechnik*. https://feucht-obsttechnik.de/en/fruit-harvesting-technology/

[7] J. A. Gil-Ribes, L. Ferguson, S. Castro-Garcia, and G. L. Blanco-Rodán, "How Agricultural Engineers Develop Mechanical Harvesters: The University Perspective," *HortTechnology*, vol. 24, no. 3, pp. 270–273, Jun. 2014, doi: 10.21273/HORTTECH.24.3.270.

[8] Y. González Castro, O. Manzano Durán, and O. García Hoya, "Critical points of change in the blackberry productive chain (Rubus glaucus Benth) in the municipality of Pamplona, Colombia," *RIDI Journal*, vol. 10, no. 1, pp. 9–22, Aug. 2019, doi: 10.19053/20278306.v10.n1.2019.10008.

[9] J. A. Sánchez Morales, M. X. Villares Jibaja, Z. Niño Ruiz, and M. B. Ruilova, "Effect of the altitudinal floor on the quality of mora (Rubus glaucus benth) in the interand region of Ecuador," *Idesia*, vol. 36, pp. 209–215, Jun. 2018, doi: 10.4067/S0718-34292018005000702.

[10] M. F. Vergara, J. Vargas, and J. F. Acuña, "Physicochemical characteristics of blackberry (Rubus glaucus Benth.) fruits from four production zones of Cundinamarca, Colombia," *Agron. Colomb.*, vol. 34, no. 3, pp. 336–345, Dec. 2016, doi: 10.15446/agron.colomb.v34n3.62755.

[11] G. A. L. Moreno, N. E. B, L. S. B. M, and C. I. M. C, "Morphological variability of wild blackberry (Rubus sp.) cultivars in the Andes of Colombia," *Colombian Journal of Horticultural Sciences*, vol. 10, no. 2, Art. no. 2, 2016, doi: 10.17584/rcch.2016v10i2.4755.

[12] S. Cancino, G. O. Cancino Escalante, and D. F. Cancino Ricketts, "Econometric model for the cultivation of blackberry (Rubus glaucus) in the municipalities of Pamplona and Chitagá, Norte de Santander, Colombia," *Aibi, Journal of Research, Management and Engineering*, vol. 8, pp. 37–43, Jan. 2020, doi: 10.15649/2346030X.752.

[13] Agronet, "Area, production and national yield by crop," *Agronet MinAgricultura*. https://www.agronet.gov.co/estadistica/Paginas/home.aspx?cod=1

[14] C. M. Grijalba Rativa, L. A. Calderón Medellín, and M. M. Pérez Trujillo, "Yield and fruit quality of andean blackberry (rubus glaucus benth), thorn and thornless, cultivated in open field conditions in Cajicá (Cundinamarca, Colombia)," *Revista Facultad de Ciencias Básicas*, vol. 6, no. 1, Art. no. 1, 2010, doi: 10.18359/rfcb.2079.

[15] M. Ramírez Gómez *et al.*, *Use of microorganisms with potential as biofertilizers in the cultivation of blackberries*. CORPOICA, 2008. Accessed: Sep. 15, 2021. [Online]. Available: https://repository.agrosavia.co/handle/20.500.12324/12735

[16] J. A. Villabona and C. L. Sandoval, "Technology for Pressure Control and Irrigation Emulation Using the Programmable Logic Control and Fully Integrated Automation Software," *Colombian journal of advanced technologies*, vol. 2, no. 32, Art. no. 32, 2018.

[17] F. Takeda, G. Krewer, C. Li, D. MacLean, and J. W. Olmstead, "Techniques for Increasing Machine Harvest Efficiency in Highbush Blueberry," *HortTechnology*, vol. 23, no. 4, pp. 430–436, Aug. 2013, doi: 10.21273/HORTTECH.23.4.430.

[18] C. L. Sandoval Rodríguez and R. Meza Pedrozo, "Application of a robust system to the position control of a pneumatic cylinder exposed to disturbances," *INGE@UAN*, vol. 2, no. 3, Art. no. 3, Dec. 2011, [Online]. Available: http://revistas.uan.edu.co/index.php/ingeuan/article/view/320

[19] R. Piyare and S. Lee, "Performance Analysis of XBee ZB Module Based Wireless Sensor Networks," *International Journal of Scientific & Engineering Research*, vol. 4, no. 4, p. 7, Apr. 2013.

[20] R. Faludi, *Building Wireless Sensor Networks: A Practical Guide to the ZigBee mesh Networking Protocol*. O'Reilly Media, Inc., 2010.

[21] A. D. Rincón-Quintero *et al.*, "Manufacture of hybrid pieces using recycled R-PET, polypropylene PP and cocoa pod husks ash CPHA, by pneumatic injection controlled with LabVIEW Software and Arduino Hardware," *IOP Conf. Ser.: Mater. Sci. Eng.*, vol. 844, p. 012054, Jun. 2020, doi: 10.1088/1757-899X/844/1/012054.

IOP Conf. Series: Materials Science and Engineering 1253 (2022) 012012 doi:10.1088/1757-899X/1253/1/012012

Probabilistic Seismic Hazard Analysis of San José de Cúcuta, Colombia

C H Flórez-Góngora[1], B S Sereno-Vargas[1], C F Lozano-Lozano[2]

[1] Grupo de Investigación en Geotecnia Ambiental GIGA, Universidad Francisco de Paula Santander, Cúcuta, Colombia. carloshumbertofg@ufps.edu.co brayanstivensv@ufps.edu.co

[2] Grupo Ambiental de Investigación Aplicada GAIA, Universidad de Santander, Bucaramanga, Colombia. ca.lozano@mail.udes.edu.co

Abstract. This work evaluated the probabilistic seismic hazard of San José de Cúcuta (Colombia) using the methodology developed by Hanks and Cornell. Seismic hazard curves were obtained using a set of ground movement prediction equation which relate the average annual rate of earthquake exceedance against maximum acceleration values, assuming that the occurrence of the events presented in the area within the time period analyzed follows a Poisson´s distribution. Additionally, the average maximum acceleration values were calculated for a set of return periods, which were compared against the effective peak ground acceleration value for the city established in the current anti-seismic design code in Colombia. For the return period of 475 years, maximum acceleration values greater than those defined in the regulation were obtained with two of the soil movement prediction equations. With the results obtained, the level of high seismic threat in which the city is located has been corroborated and it is proposed to increase the maximum horizontal acceleration for anti-seismic design code from 0.35g to 0.40g.

1. Introduction

The seismic hazard can be assessed deterministically or probabilistically. The probabilistic seismic hazard analysis (PSHA) has as its main objective to quantify the probability or exceedance rate of various levels of ground movement in a given site, for this, all related uncertainties such as magnitude are considered, the location and intensity of movement caused, taking into account all the earthquakes that may occur in the seismic sources associated with the region or study area [1] [2] [3]. To evaluate the seismic intensities associated with different probability levels, all possible scenarios are taken into account, in fact, the results reflect the contributions of all sources [4]. The PSHA was introduced by Cornell [5]. Various researchers, including Baker, suggest that, when properly implemented, it can be used to adapt to different needs that involve knowledge of seismic activity and ground movements. Atkinson and Goda quantify the intensity of ground motion using measures such as peak ground acceleration (PGA), peak ground velocity (PGV), and / or spectral accelerations for different periods of ground vibration. Subsequently, a guide was published by Hanks and Cornell [6], it explains the procedure to develop the methodology, which states that, in each seismic source, the occurrence of earthquakes in time follows a Poisson´s distribution.

However, seismic hazard studies in Colombia have been approached in a general way in order to incorporate the values of peak ground acceleration (PGA) in the earthquake resistant construction regulations. The initial threat studies were made under deterministic methodologies and the most recent ones present results under probabilistic analysis. In cities such as Ibague [7], several cities on the Pacific coast [8], Bogota [9], Villavicencio [10], Neiva [11] [12], and Tunja [13], probabilistic threat analyzes have been performed with results very close to what the studies of general threat to the country have reached. At a particular level, the city of San José de Cúcuta is located in the northeast of Colombia, an area in deep which the confluence of three tectonic plates has been identified: Nazca, South American and Caribbean, creating a complex tectonic framework with internal faults in the eastern mountain range and giving life to the so-called Bucaramanga seismic nest (BSN), which condition a high seismic threat for San José de Cúcuta. The city is also affected by faults such as Boconó and Uribante Caparo from the Andes mountain range in Venezuelan territory. In fact, the city was destroyed by a strong earthquake in 1875. The city is classified as a high seismic hazard zone in the current earthquake resistant construction code (NSR10), however, despite the seismic hazard it has not been developed the mandatory study of seismic microzoning. In this research, the seismic threat of the city of San José de Cucuta was evaluated in order to obtain the PGA values associated with return periods of 50, 475, 1000 and 2000 years, this information being of great importance to provide recommendations that ensure a better performance of the structures in the event of a future seismic crisis.

2. Materials and methods

The research was based on the PSHA methodology. The procedure was developed according to the guide proposed by Hanks and Cornell, which proposes to assume that in each seismic source the occurrences of earthquakes in time follow a Poisson´s distribution. The methodology for evaluating the city's seismic threat was carried out according to the following steps:

1. The seismogenic sources that influence the study area were identified and characterized, knowing the seismic activity that has been recorded, assuming that earthquakes of a given magnitude have the same probability of occurring anywhere within the area. The afferent area for the investigation was delimited to a rectangular area of 90,000 km2 between latitude 6.54 ° and 9.26 ° N, and longitude between -73.87 ° and -71.13 ° W, centered in the city of Cucuta. The seismic sources in the study area were classified according to the hypocentral depth of the events in cortical sources, intermediate sources and Bucaramanga Seismic Nest (BSN).

2. Three seismic catalogs were analyzed. The first is the catalog of instrumental seismic events (1993 to 2020) and historical (1644 to 1981) of the Colombian Geological Survey (SGC), the second catalog (1950 to 2020) belongs to the United States Geological Survey (USGS) and finally the catalog used in Sarmiento's undergraduate project (1644 to 1979) [14] has also been taken into account. A unified catalog was formed with 3644 events greater than or equal to 4.0 magnitude of moment Mw, since minor earthquakes do not represent a representative threat and most of the aftershocks of the events were also eliminated. For the homogenization to magnitude Mw, the relationships between magnitudes developed in Arcila et al [15] for Colombia were used. Subsequently, the subinterval of the sample of 376 years was determined within which the number of events was constant and it was assumed that this represents the complete reporting interval. According to the previous postulate, the interval of years to which a Poisson´s distribution was assumed due to the behavior of the events is given between 1950 and 2020 in a range of 70 years. The events that are located outside this interval of years were omitted and also the areas in which there were no events were subtracted from the total area analyzed, obtaining a total of 3556 events and a net area of 76344 km^2. The mean annual occurrence rate of earthquakes in the region was obtained according to the historical record of seismic events consulted.

The relationship between the magnitude and distribution of seismic events in an established period of time was defined and characterized by applying the Gutenberg-Richter´s recurrence law [16], with equation (1):

$$\log N = a - bM \tag{1}$$

Where N is the number of accumulated events with magnitude Mw greater than or equal to 4.0, which is the cut-off magnitude for the analysis, a is associated with the seismic activity of the area and b is the slope of the graph that relates the proportion of small-magnitude earthquakes versus large-magnitude events.

3. Soil movement prediction equations were used to determine the horizontal distances from the source of the earthquake, within which some measure of acceleration is equaled or exceeded for the magnitudes of

interest of the earthquakes. These equations relate distance and magnitude against peak ground acceleration. The ground motion prediction equations that were developed for areas of high seismic activity were selected, that were compatible with the type of magnitude used Mw and that the studies were published in reliable sources of information. An inventory of these empirical equations of soil motion defined over a period of years between 1964 and 2021 was compiled, summarized, and presented by Douglas [17]. The soil movement prediction equations used were those of Esteva and Villaverde and Esteva [18] [19], Joyner and Boore [20], Makropoulos and Burton and Makropoulos [21] [22], K. Sadigh reported in Joyner [23] and the Sadigh and Egan equation [24]. The chosen equations are presented in Table 1. The variables Yc, y, A, PGA refer to the value of maximum ground acceleration; M refers to the magnitude and R, d, r, r_{rup}, associate the distance of occurrence of the event in kilometers. The values of the coefficients b_1, b_2, b_3, b_4, α, β, b, h, h_1, h_2, C1, C2, C3, C4, and C5, were taken from the original equations proposed by the authors.

Table 1. Selected ground motion prediction equations

Author	Equation
Esteva and Villaverde (1973)[18] and Esteva (1974)[19]	$Yc=b_1e^{b2M}(R+b_4)^{-b_3}$
Joyner and Boore (1981)[20]	$\text{Log } y=a+\beta M-\log r+br$
Makropoulos and Burton (1985)[21] and Makropoulos (1978)[22]	$A=b_1e^{b_2M}(R+h)^{-b_3}$
K. Sadigh (1987) reported in Joyner (1988)[23]	$\ln y=a+bM+c_1(8.5-M)^{c2}+d\ln[r+h_1e^{h_2M}]$
Sadigh and Egan (1998)[24]	$\ln PGA=C_1+C_2M+C_3\ln[r_{rup}+e^{(C_4+C_5M)}]$

With the distances (R) obtained, each one was converted to a circular area (Ao) of the form πR^2, then each value was divided over the net study area (A), this relationship of areas Ao / A determines that the probability exists the city is in the zone of action of some level of maximum ground acceleration given that a seismic event occurs somewhere in the zone. Finally, the average annual rate of total exceedance was obtained for certain fixed accelerations as the sum of the products of the annual seismicity rate obtained previously by the relation of areas for each magnitude.

4. Seismic hazard curves were constructed by plotting total annual mean exceedance rates against fixed PGA values. The procedure was carried out for each attenuation equation, with which 5 seismic hazard curves were obtained for the city of San Jose de Cucuta. To express the results in terms of return period and compare them with those defined in the NSR10, the acceleration curves were constructed against fixed return periods, which were obtained as the inverse of the probabilities of exceedance of the accelerations calculated with the equation (2).

$$P = 1 - e^{-\lambda t} \quad (2)$$

Where the exceedance probability P depends on the mean annual earthquake exceedance rate λ and the analyzed time t (earthquakes per year). The PGA values and an average of them for the return periods of 50, 475, 1000 and 2000 years were obtained with each prediction equation of the movement of the soil and a percentage comparison of the value of Aa of the NSR10 with respect to the value obtained from PGA.

3. Results

The following results are presented according to the applied methodology:

3.1. Seismogenic sources and seismic activity in the region

The sources identified are cortical sources, intermediate sources and BSN, which have important effects on the seismic risk of the city of San Jose de Cucuta. The cortical sources were associated with those seismic events whose hypocentral depth is less than 70 kilometers and are related to the different geological fault systems that are present in the region. Among the most important cortical sources in the region are the Bocono Fault System (BFS) [15, 25, 26], the Uribante-Caparo Fault (UCF) [27] and the Frontal Fault System of the Eastern Cordillera (FFSEC) [28]. Intermediate sources were associated with seismic events occurring at depths greater than 70 kilometers that are not close to the NSB. Events generated at depths greater than those from cortical sources were associated with BSN [29]. Figure 1 show the epicentral location (latitude and longitude coordinates). Figure 2 shows the hypocentral location of the seismic events that occurred in the study area according to the unified catalog. The dark blue points coincide with earthquakes from the BSN, the purple points are earthquakes from intermediate sources and the red points identify earthquakes from cortical sources approximately.

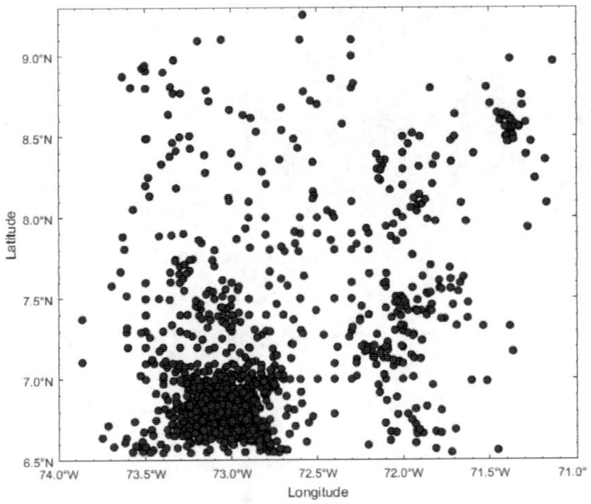

Figure 1. Epicentral location

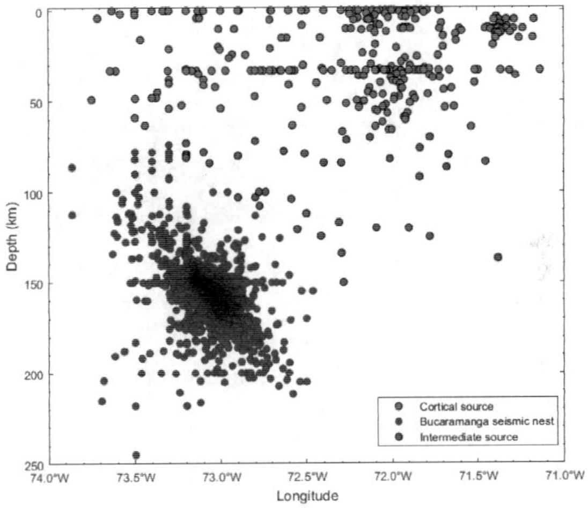

Figure 2. Hypocentral location

3.2. Seismicity rate and seismic recurrence law

3.2.1. Seismicity rate

To characterize the seismicity of the region, the frequency of seismic events with magnitudes greater than or equal to 4.0 Mw was analyzed. Figure 3 shows the frequency histogram of events that occurred in the 70-year time interval.

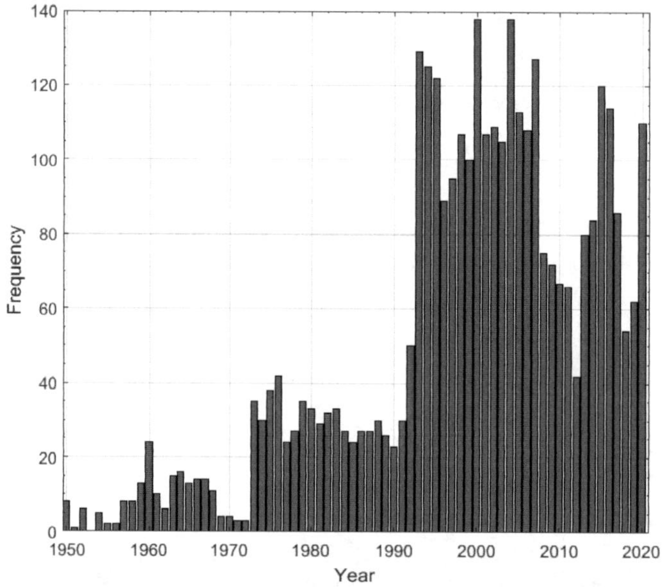

Figure 3. Histogram of the frequency of seismic events between 1950 and 2020 for an afferent area of San Jose de Cucuta of 76 334 km^2

The histogram shows an increase in the frequency of events starting in 1993, this is due to the start-up of seismological stations by the SGC in the territory and the monitoring that it began to exercise since that year. The records that precede this year are mostly registered by the USGS since 1950 and supplemented with the catalog used in Sarmiento [14]. Additionally, the mean annual earthquake occurrence rate was determined for events with a magnitude greater than or equal to 4.0 Mw. The results obtained are presented in Figure 4 for each magnitude in which at least one event has occurred.

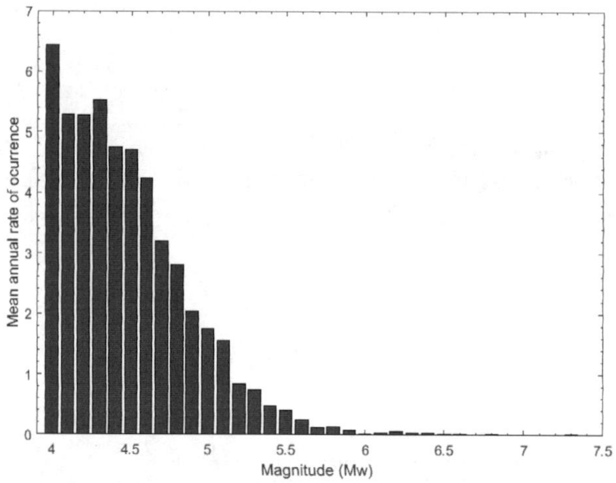

Figure 4. Average annual rate of occurrence of seismic events in the 70-year interval

In the time analyzed, a high occurrence of seismic events with magnitudes less than 5.1 Mw was evidenced, presenting values greater than 1.5 earthquakes per year, on the contrary, events greater than this magnitude have occurred with an occurrence rate of less than 1 per year. Seismic events with a magnitude greater than or equal to 6.0 Mw have occurred 15 times in the span of 70 years, which represents a total rate of 0.214285 earthquakes per year.

3.2.2. Seismic recurrence law

The occurrence of seismic events is characterized by a large generation of events with low magnitudes and few events with high magnitudes greater than 6.0 Mw, in this analysis the high seismic activity of the BSN influences which contributes a large number of events between magnitudes 4.0 to 5.1 Mw which makes the slope b = -1.2218 of the regression line greater than 1 as there are many more minor events than greater ones. The coefficient a = 8.638 associated with seismic activity is equally high for the same reason. The BSN being an area with such a particular seismic genesis has a great influence on the analysis of the available information. With a regression coefficient $R^2 = 0.9902$, the regression obtained is reliable. Figure 5 shows the seismic recurrence relationship of the sources in the studied area.

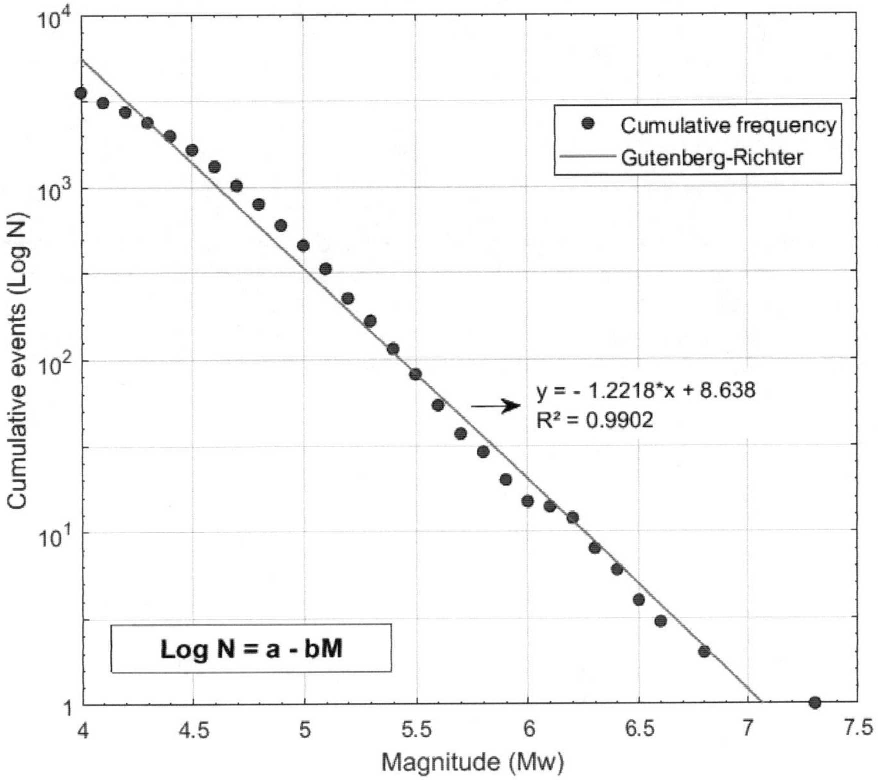

Figure 5. Seismic recurrence ratio Log N vs Mw

3.3. Application of prediction equations of ground motion

The distances within which the fixed acceleration measurements are equaled or exceeded for each magnitude of the earthquakes that have occurred were determined. To illustrate the process, the distances found against PGA values are presented with the different attenuation equations for the magnitude of 6.0 Mw in Figure 6.

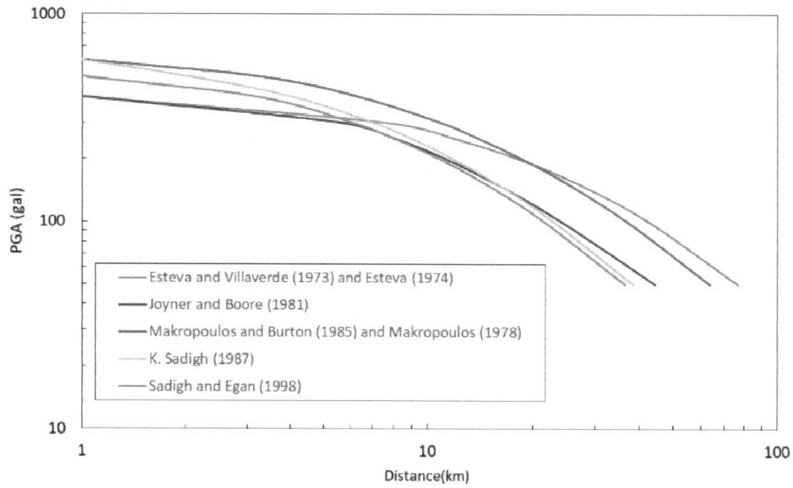

Figure 6. Attenuation ratio in earthquakes of magnitude 6.0 Mw

Using the soil movement prediction equations, the distances and area relationships were determined with which the values of the average annual rate of total exceedance per unit area were obtained for each acceleration value, this result was obtained for each magnitude seismic that has presented events in the time interval, later the summation of the contributions of each magnitude was carried out, which presents the results obtained in Table 2 for acceleration values between 50 and 700 gales.

Table 2. Average annual exceedance rate per km^2 λ

Acceleration (gal)	Mean annual rate of exceedance per km^2 λ				
	Esteva and Villaverde (1973) and Esteva (1974)	Joyner and Boore (1981)	Makropoulos and Burton (1985) and Makropoulos (1978)	K. Sadigh (1987)	Sadigh and Egan (1998)
50	1.5311517	0.9525107	1.6595735	0.5789135	0.4534584
100	0.2025019	0.1929816	0.3633704	0.1851754	0.1176591
150	0.0434516	0.0410892	0.1046350	0.0801661	0.0391241
200	0.0136531	0.0092213	0.0343294	0.0381505	0.0134581
250	0.0057695	0.0025449	0.0129575	0.0183918	0.0048019
300	0.0029556	0.0009346	0.0056094	0.0084818	0.0018562
400	0.0010226	0.0002324	0.0015363	0.0013105	0.0003742
500	0.0004132	0.0000669	0.0006027	0.0001636	0.0000976
600	0.0001922	0.0000246	0.0002803	0.0000224	0.0000216
700	0.0000961	0.0000099	0.0001395	0.0000007	0.0000017

The values obtained represent the seismic activity that occurred throughout the area in a distributed manner. It can be seen that the exceedance rate for low accelerations (50 gal to 250 gal) can have a maximum of 1.6595735 being greater than for high accelerations (300 gal to 700 gal) with a maximum of 0.029556, which is consistent with the environment. seismic present in the study area, as evidenced by the Gutenberg and Richter regression, the occurrence of seismic events with low magnitudes is greater than that of events with large magnitudes.

3.4. Seismic hazard curves

Seismic hazard curves were obtained that quantify the intensity level of the ground movement expressed as PGA values. The results for the city of San José de Cúcuta relate the average annual rate of exceedance of earthquakes per unit area in the study area against values of maximum acceleration. The procedure was performed for each attenuation equation. 5 seismic hazard curves were obtained for the city. Figure 7 shows the curves obtained for San José de Cúcuta relating the average annual rate of exceedance of earthquakes per year against the acceleration in cm/s^2. The city of San José de Cúcuta is located in a seismically active region with an occurrence rate of 50.8 earthquakes per year with a magnitude greater than or equal to 4.0 Mw in its afferent area. The region has a higher occurrence of events in earthquakes of magnitudes less than 5.1 Mw, of which most of these events between 4.0 and 5.1 Mw were associated with NSB. It is reaffirmed that the seismic threat level of the city is high.

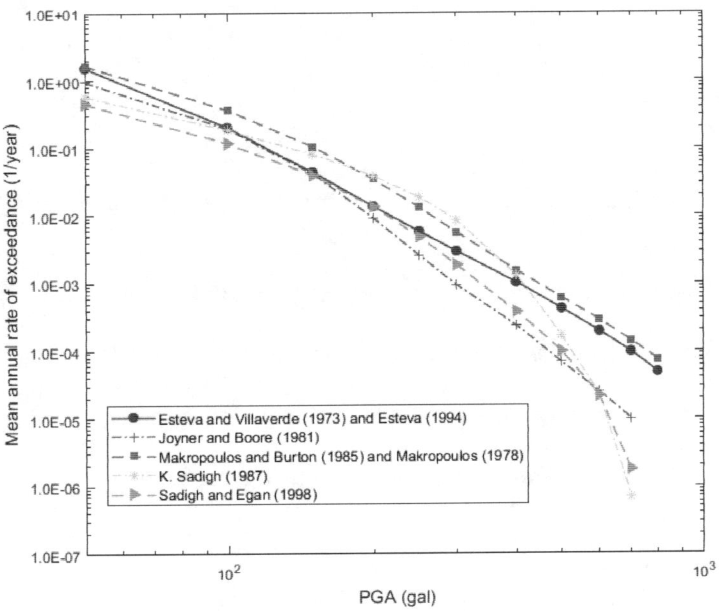

Figure 7. Seismic hazard curves for San Jose de Cucuta

The seismic hazard curves obtained reflect that the rate of exceedance of events with high accelerations of up to 700 cm / s^2 distributed throughout the study area is very low compared to an acceleration of 50 cm / s^2, in which values are reached. greater than 1 event per year with two of the curves, which is correct since the occurrence of events of low magnitude is more frequent. Compared with the hazard curve obtained in AIS (2009) [25], the trend of the curve to present more events in low accelerations with rates greater than 1 earthquake per year is preserved, however, in it is presented for accelerations of 1 gal as opposed to here. On the other hand, the exceedance rates for high accelerations are higher than those obtained with the present study.

3.5 Return periods

In Colombian regulations, it is common to express the probability of exceedance of earthquakes through the return period. In this investigation, the probable maximum accelerations that can occur given fixed return periods were obtained. Figure 8 shows the maximum probable accelerations in cm/s^2 associated with payback periods for each attenuation equation used, comparing the results with the effective peak acceleration value defined in the NSR10 for the payback period of 475 years.

IOP Conf. Series: Materials Science and Engineering 1253 (2022) 012012 doi:10.1088/1757-899X/1253/1/012012

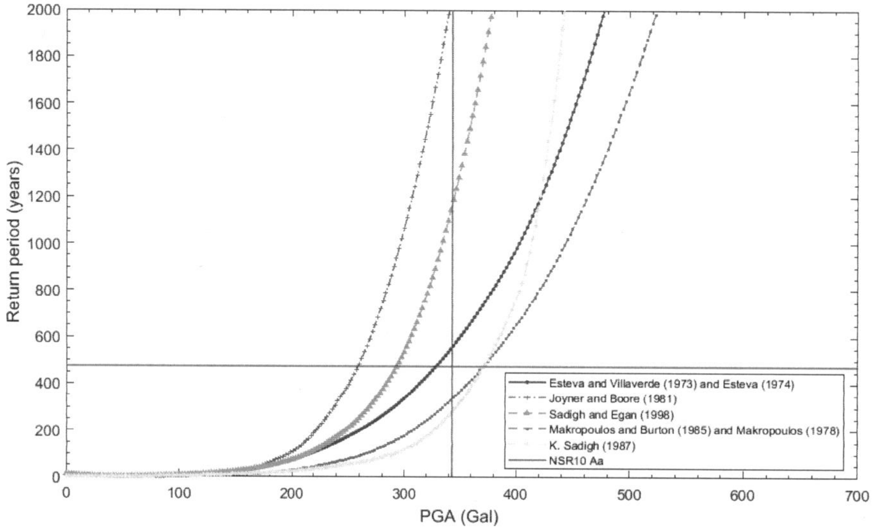

Figure 8. Acceleration Curves vs Return Period for San José Cúcuta

The maximum probable accelerations of the soil in gravity units (g) associated with the return periods of 50, 475, 1000 and 2000 years were obtained to compare the results with those defined in the NSR10. Table 3 shows the results of the maximum accelerations obtained with the methodology proposed for San José de Cúcuta.

Table 3. Maximum accelerations obtained

Return period (years)	Acceleration Aa (PGA)					
	Esteva and Villaverde (1973) and Esteva (1974)	Joyner and Boore (1981)	Makropoulos and Burton (1985) and Makropoulos (1978)	K. Sadigh (1987)	Sadigh and Egan (1998)	Mean
50	0.19 g	0.18 g	0.23 g	0.25 g	0.18 g	0.21 g
475	0.33 g	0.27 g	0.38 g	0.38 g	0.30 g	0.33 g
1000	0.41 g	0.30 g	0.45 g	0.42 g	0.36 g	0.39 g
2000	0.49 g	0.35 g	0.53 g	0.45 g	0.40 g	0.44 g

The average PGA values obtained for the city with each attenuation equation are 0.21 g, 0.33 g, 0.39 g and 0.44 g with respect to each return period. NSR10 establishes that the seismic design coefficients have an associated probability of exceedance of 10% in 50 years of useful life, which approximately corresponds to 475 years of return period, therefore, the percentage difference of the results is presented. obtained for this return period with each of the soil movement prediction equations as a form of comparison in Table 4.

Table 4. Percentage difference of Aa for return period of 475 years with respect to NSR10

	Esteva and Villaverde (1973) and	Joyner and Boore (1981)	Makropoulos and Burton (1985) and	K. Sadigh (1987)	Sadigh and Egan (1998)	Aa(g) NSR-10

	Esteva (1974)		Makropoulos (1978)			
PGA (g)	0.33	0.27	0.38	0.38	0.30	0.35
Difference	-5.7%	-22.9%	+8.6%	+8.6%	-14.3%	

Comparing the results with the values defined in the NSR10, the value of the norm of 0.35 g is 5.7% above the average value obtained with the investigation of 0.33 g in the scenario assumed as Poisson distribution. On the other hand, the highest acceleration obtained was 0.38 g, reflecting an increase of 8.6% with respect to the current value of the standard; although the NSR10 in its title A, subsection A.2.9.3.4. Literal (d) states that, when comparing the acceleration results obtained, the highest values should be used among those obtained by deterministic, probabilistic methods, spectral evaluations in rock and those established in literal A.2.2 for the city. When obtaining a PGA of 0.38 g it could be considered in a future modification of the norm, the update of the horizontal effective peak acceleration value (Aa) for Cúcuta, assigning a conservative value of 0.40g.

4. Conclusions.

The PGA values obtained with the research show an increase in the acceleration defined by the current Colombian earthquake resistant construction regulations. This would lead to the use of higher Aa values for the design of structures in the city, although in the regulation they are rounded every 0.05, the new PGA for Cúcuta must be 0.40 g in order to reduce the seismic risk for the population.

When executing the probabilistic evaluation of the seismic threat of San José de Cúcuta, it is clearly evident that more studies must be carried out on the different seismogenic sources that affect the city. Although the results show high values of maximum acceleration at rock level, site effects have not been considered, so these values could be amplified.

References

[1] Colombian Association of Earthquake Engineering (2010). *Colombian Earthquake Resistant Construction Regulation NSR-10*. Bogotá, Colombia: Colombian Association of Earthquake Engineering.

[2] Anbazhagan, P., Bajaj, K., Matharu, K., Moustafa, S. & Al-Arifi, N. (2019). Probabilistic seismic hazard analysis using the logic tree approach – Patna district (India). *Natural Hazards Earth System Sciences*, 19, 2097–2115. DOI: https://doi.org/10.5194/nhess-19-2097-2019

[3] Baker, J. (2015). Introduction to Probabilistic Seismic Hazard Analysis. *White Paper Version 2.1.*

[4] Atkinson, G. & Goda, K. (2013). Probabilistic seismic hazard analysis of civil infrastructure. In: *Handbook of Seismic Risk Analysis and Management of Civil Infrastructure Systems*. S. Tesfamariam, K. Goda (Cambridge: Woodhead Publishing Ltd), 3-28, doi: 10.1533/9780857098986.1.3

[5] Cornell, C. (1968). Engineering Seismic Risk Analysis. *Bulletin of the Seismological Society of America*, 58 (5), 1583-1606.

[6] Hanks, T. & Cornell C. (1994) Probabilistic Seismic Hazard Analysis: A Beginner's Guide. In *Proceedings of the. Fifth Symposium on Current Issues Related to Nuclear Power Plant Structures, Equipment and Piping*, North Carolina State University, Raleigh, 1-17.

[7] Romero, M. (2007). *Probabilistic evaluation of the regional seismic hazard for the city of Ibagué*. Thesys, University of La Salle. Bogotá, Colombia.

[8] Martínez, J. (2007). *Probabilistic evaluation of the seismic hazard of the main urban centers of the Colombian Pacific Coast (Quibdó, Buenaventura and Tumaco)*. Thesys, University of La Salle. Bogotá, Colombia.

[9] Alfaro, A. (2007). Uncertainties in the Evaluation of the seismic hazard for Bogotá – Colombia. *Épsilon*, 9(7), 73-87.

[10] Ramírez, M. (2007). *Probabilistic evaluation of the regional seismic hazard for the city of Villavicencio.* Thesys, University of La Salle. Bogotá, Colombia.

[11] Bohórquez, M. (2007). *Probabilistic evaluation of the regional seismic hazard for Neiva*. Thesys., University of La Salle, Bogotá, Colombia.

[12] Alfaro, A. (2013). Probabilistic and deterministic evaluation of the seismic hazard of Neiva (Colombia) using the new generation of attenuation equations. *Épsilon*, 20(8), 145-158.

[13] Molina, G. (2008). *Probabilistic evaluation of regional seismic hazard for Tunja*. Thesys, University of La Salle. Bogotá, Colombia.

[14] Sarmiento, M. (2004). *Internship in the administrative department of municipal planning for the elaboration of the methodological base of the preliminary seismic microzonation of San José de Cúcuta*. Thesys, University Francisco de Paula Santander. Cúcuta, Colombia.

[15] Arcila, M., García, J., Montejo, J., Eraso, J., Valcarcel, J., Mora, M., Vigano, D., Pagani, M. y Díaz, F. (2020). *National seismic hazard model for Colombia* Bogotá: Colombian Geological Services and Global Earthquake Model Foundation. https://doi.org/10.32685/9789585279469

[16] Gutenberg, B. & Richter, C. (1944). Frequency of earthquakes in California. *Bulletin of the Seismological Society of America*, 34,185-188.

[17] Douglas, J. (2021). *Ground motion prediction equations 1964-2021.* Department of Civil and Environmental Engineering, University of Strathclyde. United Kingdom.

[18] Esteva, L., and Villaverde, R. (1973). *Seismic risk, design spectra and structural reliability.* In Proceedings of Fifth World Conference on Earthquake Engineering, (2), 2586-2596.

[19] Esteva, L. (1974). *Geology and probability in the assessment of seismic risk.* In: Proceedings of 2nd International Conference, Association of Engineering Geologists, Sao Paulo, Brazil.

[20] Joyner, W. & Boore, D. (1981). Peak horizontal acceleration and velocity from strong-motion records including records from the 1979 Imperial Valley, California, earthquake. *Bulletin of the Seismological Society of America*, 71(6), 2011-2038.

[21] Makropoulos, K. and Burton P. (1985). Seismic hazard in Greece. II. Ground acceleration. *Tectonophysics*, 117, 259-294.

[22] Makropoulos, K. (1978). *The statistics of large earthquake magnitude and an evaluation of Greek seismicity.* PhD thesis, University of Edinburgh, UK.

[23] Joyner, W. and Boore, D. (1988). *Measurement, characterization, and prediction of strong ground motion.* In Proceedings of Earthquake Engineering & Soil Dynamics II, Geotechnical Division, ASCE, P.43-102.

[24] Sadigh, R. & Egan, J. (1998). *Updated relationships for horizontal peak ground velocity and peak ground displacement for shallow crustal earthquakes*. In Proceedings of the Sixth U.S. National Conference on Earthquake Engineering.

[25] Diederix, H., Torres, E., Hernández, C. & Bohórquez, O. (2009). *Tectonic and morphodynamic evolution during the Quaternary in the area of Cúcuta and surroundings, Norte de Santander.* Report, Bogotá, Colombia: INGEOMINAS.

[26] Londoño, J., Bohorquez, O., & Ospina, L. (2010). 3D seismic tomography of the Cúcuta sector, Colombia. *Boletín de Geología*, 32(1), 107-124.

[27] Colombian Association of Earthquake Engineering. (2009). *General study of seismic hazard in Colombia 2009*. Bogotá, Colombia: Colombian Association of Seismic Engineering.

[28] Paris, G. & Romero, J. (1993). *Active faults in Colombia*. Bogotá, Colombia: INGEOMINAS.

[29] Sepúlveda, F., & Cabrera, F. (2018). 3D seismic tomography of the Bucaramanga seismic nest (Colombia). *Bulletin of Geology*, 40(2), 15-33.

Machine Learning algorithms to predict desertion in the faculty of Engineering Sciences at the Corporación Universitaria Antonio José de Sucre.

C M Pineda-Pertuz[1], Y. Martinez[1], I. Diaz[1]

[1]Corporación Universitaria Antonio José de Sucre (Colombia).

docente_ingenieria1@corposucre.edu.co

Abstract. Desertion can be understood as the withdrawal and subsequent abandonment of school activity by students, due to different family, economic, social and other factors. All educational levels are affected by this scourge, being the university one of the most affected, especially during this time of pandemic caused by the SARS-CoV 19 virus. Due to the complexity of this problem, and the great impact it has generated at the educational level, many universities have proposed different intervention strategies to reduce dropout rates. The difficulty is that many of these strategies lack effectiveness, since they do not take into account the different causes, which are different for each case. On the other hand, it is important to have accurate and reliable information to determine the population in order to identify possible cases of dropout, and to take preventive actions to reduce the student dropout rate. In this sense, the purpose of this study is to analyze different predictive algorithms based on Machine Learning that enable early detection of possible cases of desertion in the Faculty of Engineering at the Corporación Universitaria Antonio José De Sucre. Among such algorithms we can mention decision trees, logistic regression and support vector machines. These will be trained with historical data and then tested to determine their performance, to finally choose the most appropriate within the context of the current problem.

Keywords: desertion, Machine Learning, model, prediction, algorithm.

1. Introduction

According to the Ministry of National Education, dropout can be understood as the abandonment of the school system by students, caused by a combination of factors generated both within the system and in social, family, individual and environmental contexts. [1]

University dropout is one of the causes that lead to low performance in academic programs, motivating students entering university careers to abandon their studies.

According to a study conducted by the World Bank, Colombia is the second country in Latin America with the second highest university dropout rate, with 42% of students withdrawing from institutions in the first semesters. Thus, the country is only preceded by Bolivia, where the rate reaches 48%. Also, according to the OECD, only 22% of the population between 25 and 34 years of age has a university degree, while the average for the organization's member countries is 38%. [2].
Currently, some institutions of higher education have been seriously affected by the problem of desertion, especially in recent times due to the emergency caused by the covid-19, among these institutions of higher education we find the Antonio José de Sucre University Corporation. Due to the wide variety of programs offered by the institution, there are many students who are motivated to be formed in this, but later they are in the difficult situation of abandoning their studies for various reasons ranging from academic, economic and social. The institution, aware of this problem, has a Student Permanence program where activities and spaces oriented to the prevention of desertion are promoted, given that, if it is not given due attention, it would be evident a decrease of important resources for enrollment, as well as an increase in social problems that undoubtedly contribute to a greater backwardness in the municipality and the region.
Educational institutions, aware of the increasing demands and challenges in this area, have found it necessary to integrate technology and innovation in their processes in order to offer a more efficient service. Specifically, artificial intelligence is one of the disciplines that has aroused more interest in recent times, being the vehicle for a wide variety of solutions, through its many techniques and tools that have contributed to repair different problems within universities and the business sector in general.
Therefore, many institutions have implemented models using different strategies such as the implementation of machine learning algorithms to prevent cases of attrition.
Machine learning, also known as Machine Learning, is a branch of artificial intelligence that aims to develop techniques that allow machines to learn on their own [3].
Regarding the use of these algorithms to predict student dropout using personal and academic information of students. The authors [4] developed a predictive system to detect the student likely to drop out. Using Microsoft SQL Server Analysis Services tool, they created a prediction model based on discrete and continuous attributes, and a classification algorithm called decision tree to find the students with high dropout rate.
Similarly, [5] propose the construction of a model to predict the probability of students dropping out of their university academic program, also using classification techniques based on decision trees. Similarly, [6] propose a neural network-based model to predict engineering students at risk of dropping out based on their socioeconomic, academic and personal profiles, based on the CRISP-DM methodology.
Therefore, the objective of this research is the analysis of the performance of several machine learning algorithms, suitable for making predictions, in order to facilitate and support the processes of identification of potential dropouts in the Faculty of Engineering Sciences at the Corporación Universitaria Antonio José de Sucre. This process was based on an analytical study of student data from the last 10 academic periods. This information was retrieved from the institutional academic system and organized into a data set (.csv file), which was then duly preprocessed using Python libraries such as Pandas, Numpy and Scikit-learn. When the data was already fit. That is, without any inconsistencies and in numerical format, they were passed to the predictive algorithms in question, to be trained and then evaluated, so that it is possible to know how well they performed, the latter through the application of metrics such as: accuracy, precision, recall and f1. Finally, once the response given by each algorithm was known, it is determined which one offers the best result in terms of predicting a possible defector.

2. Methodology

This research has a quantitative approach and it will address dropout as the main variable and all the variables that have an impact on it. As a result of this research, it will seek to support the work done by institutional welfare for the Faculty of Engineering Sciences of the Corporación Universitaria Antonio José de Sucre around the timely prevention of desertion and thus mitigate the rates that have been presented in recent years. The development of this research will be approached through the agile SCRUM methodology, where teamwork, feedback with the client, follow-up meetings and adaptation to change are the most important.

In addition, the requirements are captured in user stories with the help of the people involved in the project. The user stories serve as an open questionnaire where the needs to be solved during the development of the work are established.

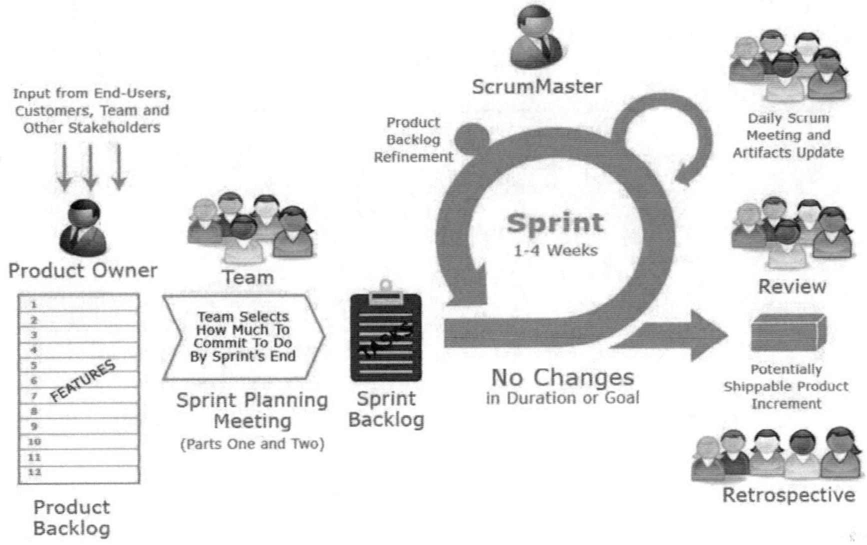

Figure 1. Elements of the Scrum Methodology

Scrum development is fragmented in small periods of work time, in which work is done in a concentrated way, called Sprint. These Sprint, allow a high pace of work, in part because the team is highly qualified, and with sufficient authorizations to make relatively important decisions on the fly, eliminating waiting times that do not contribute anything [7].

The general process using the Scrum methodology can be described in the following 3 stages:

Planning:

For this stage, special attention was paid to the gathering of user requirements through the implementation of user stories, as well as defining the roles and people responsible for carrying out each user story, and also defining the dates of the project deliverables.

Springs Development:

At this stage, each of the user stories was developed by each person in charge. As each story was developed, the programmer marked it as finished in an online dashboard created with the Jira software.

Springs Evaluation:

Finally, at this stage, the validation of each one of the Springs was carried out, where the client (teacher) checked that the proposed functionalities complied with the established requirements.

3. Proposed solution

As part of these user stories, we identified the variables that have an impact on attrition, which were classified according to different factors, as illustrated below:

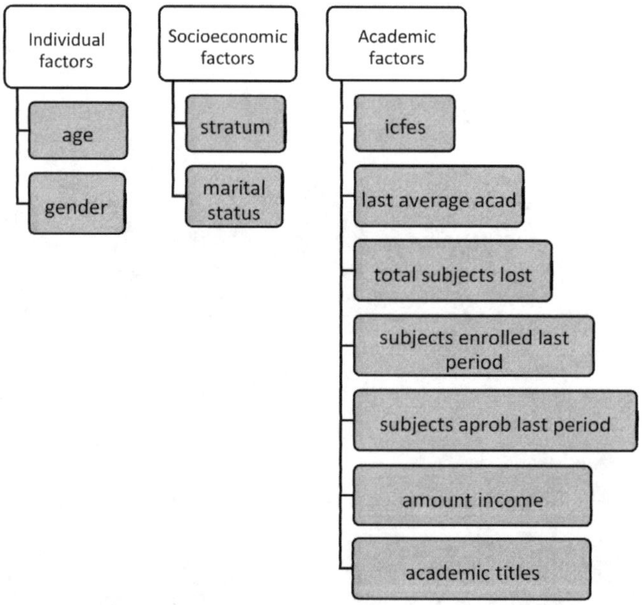

Figure 2. Classification of variables

With the above variables and the historical data obtained through the institution's academic information system, a Pandas dataframe was consolidated with 11 columns plus the target variable and 1171 rows, to which the typical stages of a Machine Learning process were applied, which are illustrated below:

IOP Conf. Series: Materials Science and Engineering 1253 (2022) 012013 doi:10.1088/1757-899X/1253/1/012013

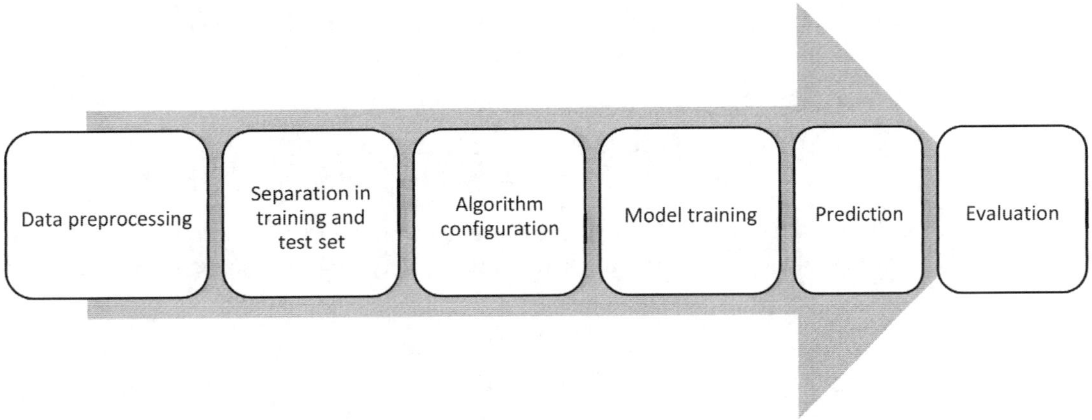

Figure 3. Machine learning process

The algorithm configuration phase was the only one that changed in each test performed, since 6 classification algorithms were used: Decision Tree, Random Forest, Adaptive Boosting, Gradient Boosting, Logistic Regression and Support Vector Machines.

Data preprocessing: This is one of the most important stages, since it involves tasks such as cleaning and transformation of the data so that they are in a suitable form and can be used by the machine learning algorithm. It is a crucial stage, since these algorithms work only with numerical data and, if possible, on the same scale. In addition, missing or null data must also be taken into account as they can cause the algorithm to operate poorly.

Separation of training and testing: In machine learning, the dataset is usually separated into two subsets called training and testing. The first one is used to train and estimate the model parameters. On the other hand, the second is used to make predictions and test the model with data different from the training data to see if it yields the expected results.

Algorithm configuration: In this stage, basically an instance of the algorithm to be used is created and the so-called hyperparameters for that algorithm are defined with appropriate values that the data scientist or programmer must adjust with care. Hyperparameters differ from parameters in that the latter are defined internally by the model during training, with the user having no direct involvement in the generation of their values. Some hyperparameters regularly used in machine learning algorithms are *the number of epochs and the learning rate*.

Model training: It consists of providing the object or instance of the machine learning algorithm with a set of training data so that it can learn, thus estimating the parameters of the learning model.

Prediction: Once the model has been generated, its level of prediction can be tested by passing samples from the test set. The result of the prediction on a sample is a continuous or discrete value that should be as close as possible to the expected value.

Evaluation: Evaluation is the process of numerically determining how effective our machine learning model was. The metrics used in this activity are described below:

- Accuracy: Measures the percentage of cases that the model has been correct. It is the ratio of correct predictions to the total number of predictions.

$$Accuracy = \frac{VP + VN}{VP + VN + FP + FN}$$

- Precision: Is the ratio of predictions correctly classified as positive to the total number of correct predictions made (both positive and negative). (both positive and negative).

$$Precision = \frac{VP}{+VN} VP$$

- Recall: Corresponds to the number of elements correctly identified as positive out of the total number of true positives.

$$Recall = \frac{VP}{VP + FN}$$

- F1: This metric allows combining the accuracy and sensitivity measures into a single value. This can be useful in cases where it is desired to compare the combined performance of these metrics.

$$F1 = 2 * \frac{precision * recall}{precision + recall}$$

4. Results

The evaluation of the 6 machine learning models generated produced the following results.

Table 1. Algorithm evaluation

Algorithm	Metrics			
	Accuracy	Precision	Recall	F1
Decision tree	0,86	0,91	0,87	0,89
Random forest	0,77	0,80	0,87	0,83
Adaptive Boosting	0,83	0,84	0,91	0,87
Gradient Boosting	0,86	0,85	0,96	0,90
Logistic regression	0,86	0,85	0,96	0,90

Support vector machines (SVM)	0,63	0,65	0,96	0,77

The above table shows that the decision tree algorithm is the one that shows the best accuracy and precision, which makes it one of the most used algorithms for this type of problem such as attrition analysis.

In addition, a decision tree is a classification algorithm that can also be used for regression. Also, decision trees are suitable when you want to show how a decision-based process flows, and they allow you to manipulate multidimensional data with good accuracy. They can be used on non-normalized data, since each attribute is handled independently and there is no restraint with respect to its scale, and as if that were not enough, they are useful to know which are the most influential characteristics of a data set, as we can see in the following image. Where it is clear that the variables *amount income, last_average_acad, age, academic_titles and stratum* are the most important determinants of attrition.

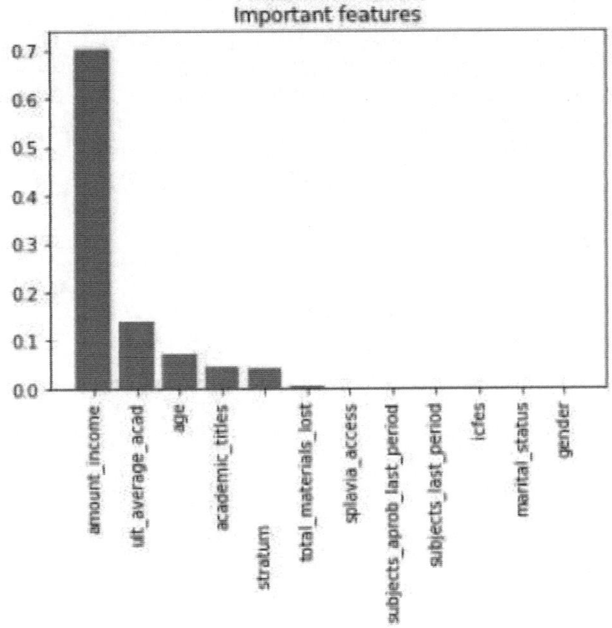

Figure 4. Variables most influential in attrition

5. Conclusions

It was possible to evaluate different machine learning algorithms, and it was possible to verify that the decision tree algorithm is the one that provides the best performance on the collected data set.

The use of an agile programming methodology Scrum has allowed to maintain a permanent communication between all those involved in the project including, of course, the clients, thus ensuring a better control and monitoring of the project, since tools of the work team domain have

IOP Conf. Series: Materials Science and Engineering 1253 (2022) 012013 doi:10.1088/1757-899X/1253/1/012013

been used to visualize and have a balance of the pending tasks, under development and solved of each Spring and also acceptance tests have been performed at the end of each of these development periods.

The main recommendation is to increase the number of instances of the dataset so that the selected machine learning model can have a much higher performance. Likewise, more tests should be performed taking into account different combinations of hyperparameters in order to improve the quality of the results obtained.

Bibliography

[1] Mineducacion, «Minieducacion,» 2021. [On line]. Available: https://www.mineducacion.gov.co/1621/article-82745.html.

[2] «Sectorial,» 14 February 2020. Available: https://www.sectorial.co/articulos2 especiales/item/296882-deserci%C3%B3n-universitaria,-%C2%BFmoda-en-las-nuevas] generaciones-o-limitantes-de-la-educaci%C3%B3n-superior.

[3] G. J. Giner, « Introduction to Machine Learning," School of Business and Management,», 3 2018.

[4] R. A. M. L. M. A. G. J. L. T. I. A. A. G. Hernández, Noviembre 2016. [En línea]. Available: https://www.researchgate.net/publication/311855420_Comparative_Study_of_Algorithms_to_P] redict_the_Desertion_in_the_Students_at_the_ITSM-Mexico.

[5] W. G. R. S. Blanca Cuji, « Predictive Model of Student Dropout Based on Decision Trees," Espacios Magazine, p. 17, 2017.

[6] H. J. C. B. D. D. S. G. J. y. Z. G. A. Senturion Cardeña, « Neural Networks to Predict Academic Dropout in Engineering, Technology and Education,», 2020.

[7] J. S. Rial Huerta, «Deposito De Investigacion Universida de Sevilla,» 2019. [On line]. Available: https://idus.us.es/handle/11441/87251.

Characterization of learning profiles, according to perception channels, in systems engineering students.

I J Romero [1], A D Morales[2], S A Sánchez [3]

[1]Corporación Unificada Nacional, Colombia,
Engineering School, AXON Research Group
[2]Corporación Universitaria Remington, Sincelejo, Colombia,
Faculty of Engineering, INGENIAR Research Group
[3]Corporación Universitaria Antonio José de Sucre, Sincelejo, Colombia,
Faculty of Engineering Science, GINTEING Research Group

ingrid_romero@cun.edu.co

Abstract. This article refers to characterize the predominant cognitive preferences or also called predominant perception channels of first semester students of Systems Engineering, virtual modality, associated to a subject in the area of Computer Science and Technology. The study was carried out with the purpose of providing teachers with a valuable informative tool that serves as a basis for proposing methodological innovations that contribute to face the difficulties of teaching concepts related to computer architecture, taking into account the differences, of individuals, associated to sensory perception, according to the CHAEA model, which has been validated by several authors. The research was descriptive, showing that the predominant profiles in the groups of students who participated in the research are those corresponding to the Theoretical, Pragmatic profile, the former being the predominant one; with a lower tendency for the Reflective style. It is concluded that there is a tendency of diversity towards the 2 channels of perception in which a higher level of preference was observed on the part of the students; therefore, it is suggested that the teaching strategies required for these students be varied according to the predominance found.

Keywords: Technology, CHAEA, education

1. Introduction

Educational institutions, especially universities, in Latin America have been participants in an increase in the number of students, which has brought with it the need for changes in their structure at the functional and organizational level, facing the continuous changes that this sector demands, which is why one of its challenges is the need to adapt to a new progressive social reality and in parallel rooted in technology and the use of Information Technology -IT for educational action [1]. In the first decade of the 21st century, globalization allows the increase of communications and consequently the transmission of information and knowledge in a vertiginous way, bringing with it at the same time the requirement for higher education institutions to provide a quality process that supports students and teachers in the learning and teaching processes, so that meaningful learning is achieved and students can reach appropriate learning levels to put into practice the knowledge acquired in order to contribute to solve problems of their environment and the continuous demands of the market [2], [3]. Therefore, it is important to highlight the active role of students and teachers in the processes that higher education institutions must carry out in order to achieve the quality of the educational service.

The profile of university students has changed as time has gone by, and this is how the student body in Latin America has incorporated the female gender, students from the interior of the countries, professional students, students who are in the labor field, among others, therefore, they require an education mediated by technology, an education that allows them to have that distance learning space with flexible times and that can be adapted to their lifestyle, so that they can obtain quality training without interrupting the daily process of the demands of their daily lives, this is how educational institutions have ventured into a different approach to traditional education, providing in addition to the face-to-face modality, virtual education options [4], [5].

Virtual education has the particularity of having spaces of student diversity, i.e. a group of students can come from different types of areas of knowledge, contexts, ages, living conditions, in view of which, the Higher Education Institutions - EIS in Colombia, face great challenges that deserve to be confronted with strategies that contribute to teachers to facilitate that students can acquire knowledge in the best possible way, taking into account that each of them has a different environment, expectations and a way of learning according to their condition [6], [7].Through virtual education, students learn to be independent, autonomous, self-taught, self-manager of their own progress, accompanied in the process, by a teacher guidance, thus aiming to achieve goals present in other modalities, such as having accessible a suitable and contextualized quality learning environment; in this sense, knowledge can be assimilated in a way that is compatible with their particularities. Faced with this situation it is important to highlight that, although the literature shows us that teachers are decisive in achieving learning objectives, it is still not clear how they can be provided with elements or tools for them to achieve it [8], [9] [10], is thus how one way is contributing Is contributing to their professional work with elements that contribute to their strategic analysis in the face of a population facing an increasingly complex and variant social reality. It is to highlight the importance of teachers being able to know the characteristics of the students they guide, to be able to analyze, establish, share, replicate and adapt strategies in favor of generating in the classrooms an environment according to these challenges.

From this perspective, the purpose of this study focuses on the characterization of the predominant learning profiles of first semester university students of systems engineering at CUN, based on the channels of perception of preference, given the need to set precedents on the learning styles presented by students who opt for the virtual modality; an important aspect to strengthen the quality of virtual education, by having relevant inputs for improvements in the teaching-learning processes of the institution, in accordance with its pedagogical model. In this sense, the study provides a perspective that glimpses the preferences of students in terms of how they perceive information, and the predominant learning profile, without discarding preferences at the intermediate level. The study contemplates that such information is relevant to generate contributions to the knowledge that teachers should have regarding how their students learn in a better way, so that it becomes a useful tool for future processes of analysis and application of methodological strategies, according to the student context, knowing that there is little specialized literature on the subject.

To fulfill this purpose 308 first semester students of the systems engineering program virtual modality who took the subject computer science and technological convergence, through the Moodle platform, during the first and second academic period of the year 2021. The students completed the CHAEA questionnaire of the authors Alonso Gallego and Honey [11] composed of 80 items to identify in the students the categories active reflective theoretical and pragmatic, the instrument is based on prominent authors in the teaching and learning processes based on learning profiles (AP) [12], [13] [14], [15], [16] [17]; being possible, in a wide spectrum, to characterize students and recognize their preferences to perceive information, their multiple ways of learning and the predominant learning profiles.

This research is referentially and theoretically based on the concepts of perception channels, profiles and learning styles, approached by several authors in favor of strengthening teaching and learning processes. Cacheiro [18], states that teaching-learning processes need resources that help to realize cognitive, procedural and attitudinal knowledge. According to Zapata [19], educational resources based on virtual technologies should be taken as an alternative to complementary education in the teaching of intellectual contributions in schools of education, with educational material coming from virtuality, allowing to complement traditional education.

The research work of García [20], shows us virtuality as a model of relevance in academic training, showing us the contribution it has in education, and motivating the use of emerging technologies in education. The author states that education in the digital world requires different educational resources, which should promote openness, initiative, autonomy, interaction and "community" learning, encouraging critical thinking to contrast and solve problems. The pandemic shows that to prepare for the digital world, teaching must be flexible, active, innovative, interconnected and personalized to respond to individual needs and interests.
A study conducted by Claudia Dania and Susana Marchisio [21] aimed to determine the main cognitive preferences of students in the information systems engineering career at the Latin American Education Center in Argentina. The research focused on proposing innovative methodologies that incorporate multimedia in a way that allows facing the challenges of teaching technical concepts and obtaining positive results, taking into account the student as a unique being who has a learning profile associated with sensory perception, according to the validated model. The research made it possible to deduce that these individual differences related to sensory perception should be considered when teaching technical concepts related to computer science and computer programming.

Each person has a particular way of perceiving his environment, characterized by being unique, and thus has a specific way of learning, which applies to various aspects of his life and is related to his experiences, skills and way of learning [22.] In order to obtain an optimal learning state, brain and body integration is required, so that information is perceived by the senses and these transmit it to the brain. Sensory preferences are also called perception channels since information is perceived through them [23].
From a neurolinguistic perspective, it is considered that the channels can be classified into areas or dimensions, mainly visual, auditory and kinesthetic. Students usually feel closer or more familiar with one perceptual channel, although they may be able to use, to a certain degree, several of them in learning. A predominant perception channel makes it possible to identify the predominant learning profile, and is therefore very useful for diagnostic studies and analyses. It is convenient for teachers to take into account that in a classroom there will always be students who have developed a learning channel or combinations of several, in other words, possess a predominant learning profile and others at a lower level [24].
Detecting the learning profile will help to truly understand the potential that is waiting to be activated in each individual, which is a useful element when generating meaningful learning, using various forms and speeds during the formative process, as stated by Aguilera [25], López [26], and Gutiérrez et al. [27].
Conceptually, learning styles are defined, in terms of personality and intelligence, as personal variables that explain the different ways in which an individual approaches, plans and responds to the challenges and demands of learning, manifesting themselves through a behavioral pattern[28], [29]. Milgram, Dunn and Price [30] relate learning styles to "the conditions under which each person begins to concentrate, process, internalize and retain new and difficult information, as well as the

skills to do so", while Keefe [31], conceives them as "cognitive, affective and physiological traits that serve as relatively stable indicators of how learners perceive, interact and respond to their learning environments". In contrast to the above, Velasco [32] states that learning styles are defined as "a set of biological, social, motivational and environmental characteristics that an individual develops from information..., to perceive and process it, ..., construct concepts, categories and solve problems. These together establish their learning preferences and define their cognitive potential". Being consistent with the above, it is concluded that, learning styles are directly related to learning preferences, and that, in turn, the latter with the predominant learning profile when a greater preference tendency is identified. The literature shows that there is a certain consensus among the authors at the moment of conceiving that they analyze stable traits, although it is worth highlighting that they can be susceptible to changes and improvements in the course of time, taking into account the diverse situations of a subject; that, when taken into account at the moment of designing a learning environment, students could learn more effectively, and obtain significant learning.

In this study, from the theoretical approach, the literature of those authors, theories and instruments that have made significant contributions to the area of learning styles and profiles and that are efficiently used to support all processes involving the acquisition of knowledge are considered. It is valid to clarify that those contributions that meet the criteria of validity and reliability will be addressed.

According to the above, we will start with the oldest referent and end with the most recent and updated ones, without ignoring that the topic of learning style is constantly renewed as research on this subject progresses.
In this order of ideas, Honey and Mumford [33] are based on Kolb's theory for the elaboration of the learning styles questionnaire (LSQ), this test has a business approach and seeks to determine what factors influence 2 people in the same situation to have different responses to the learning process, although this instrument has been frequently used, some limitations are evident regarding its application, since it was elaborated for managerial and organizational contexts.

McCarthy [34] proposes a theory of learning styles based on the studies carried out by Kolb in relation to the cerebral hemispheres and their implication in the way we learn. For this purpose, the author proposes a learning type meter called LTM (Learning Type Measure), which in general terms is a 26-point scale that measures four basic types of learning styles through the assessment of individual preferences: imaginative, analytical, common sense and dynamic. It is important to emphasize that this instrument has the necessary validity and has therefore been used in several studies.
Another fundamental contribution in the area of learning style was made by Fleming and Mills [35], who developed an instrument to determine, from a sensory point of view, the preference of students when processing information. The authors speak of sensory preferences in that students constantly receive information through the senses and that the brain selects some of this information and ignores the rest. The information is selected according to their interests, without neglecting how the information is received. Their model takes the name VARK from the acronym for the sensory modalities they identified: Visual (V), Auditory (A), Reading (R), Kinesthetic (K).

The Learning Styles Questionnaire CHAEA by Alonso, Gallego and Honey [11], which adopts the LSQ questionnaire to the academic environment and to the Spanish language, this instrument consists of 80 items and describes the styles as active, reflective, theoretical and pragmatic.
Likewise, for Dunn, Dunn and Price [14], learning style is: "the way in which 18 different elements (later increased to 21), which come from 4 basic stimuli, affect a person's ability to absorb and retain". Taking into account this definition, it can be inferred that this definition proposed by the author described is based on the same proposal made by the previous author; however, it is criticized because it does not incorporate the element of intelligence, which is fundamental in the study of Learning

Styles. Therefore, the CHAEA questionnaire is selected for this study because it is directly related to the identification of predominant learning profiles in students, taking into account their preferences when perceiving information and the channels of sensory perception, with the intention of having a broad spectrum so that the resulting information contributes to the diagnosis of the student group and consequently to future designs of didactic material for the teaching of subjects related to Computer Science and computer architecture, in the first instance, since this information can be analyzed for any subject related to computer science and therefore the Information and Communication Technologies, in the first semester of systems engineering in the virtual modality of the CUN.

It is relevant to mention that, although differences are perceived between the theoretical foundations that accompany other instruments, with those of CHAEA, it is noteworthy that, at a general level, these theories converge in the purpose of understanding the learning process, and how the way of perceiving information sets the tone to establish channels of perception, learning styles and profiles [21], so that identifying them is a strategy to generate adjustments in the didactic planning of the subjects, oriented to potentiate the innate abilities of the students and that, according to Quiñones [36], it is possible to contribute to the development of a learning style based on the perception channel with which the student interprets his environment. In this sense, the research effort addressed in this area of knowledge is oriented to the classroom experience becoming a platform to maximize and optimize the potentialities of the individual so that his or her qualities drive him or her towards achieving better learning results.

2. Materials and Methods

2.1 Methodological design

This study has a quantitative focus using the descriptive method under a non-experimental design [37]. The data are obtained from reality, the sample consisted of a total of 308 students who took the subject of Computer Science and Technological Convergence, through the Moodle platform during the first and second academic period of the year 2021; enrolled in the Systems Engineering program in the distance mode at the National Unified Corporation - CUN. It is important to highlight that the age for students in the modality ranged from 17 to 51 years old, with an average of 27.77 ± 0.22 years. Most of them came from families of socioeconomic stratum three, followed by socioeconomic stratum two and one.

The statistical analysis of the study applied descriptive statistics with the support of SPSS software and a database in Microsoft Excel, and, in addition, univariate, descriptive analyses of mean and standard deviation were performed. The above, with the data collected from the application of the CHAEA questionnaire [11]. Regarding reliability, CHAEA has numerous studies that have determined good reliability values [6], [10], [21],[36]. In this sense, report reliability indexes of 0.627 for the active style, 0.725 for the reflective, 0.658 for the theoretical and 0.588 for the pragmatic; referring to the validity of the questionnaire they show that 15 factors explain 40% of the total variance.

3. Results

As a result of the study, we obtained the characterization of the predominant or most used perception channel by students for learning, called learning profile (PA), followed by the second in terms of use and preference, called complementary (PAC). All students filled out the instrument, and the results are shown in Figure 1 and Table 1

Figure 1.

Learning profile according to level of preference

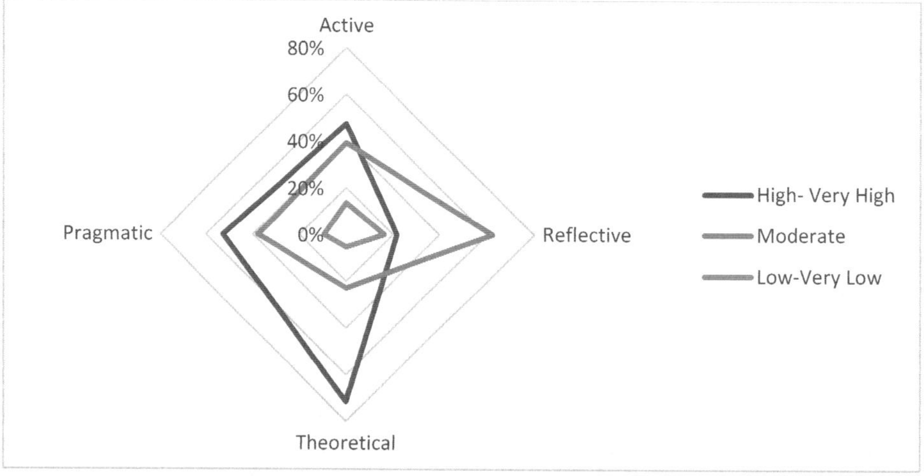

Comparing the levels of the learning preference channels, based on learning styles and the CHAEA scale, it was shown that the students have an active and reflective profile. Likewise, the profile was determined, with a predominance of the theoretical-pragmatic tendency, given that, although the most predominant was the theoretical one, it was evidenced that the pragmatic one follows it in the tendency.

Table 1.
Valor Percentage value and frequency of each perception channel

Predominance/ Learning	Frequency (F). Active	Percentage (P)	P. accumulated	F. Reflective	Percentage (P)	P. accumulated	F. Theoretical	Percentage (P)	P. accumulated	F. Pragmatic	Percentage (P)	P. accumulated
Frequency (F) High-Very High	145	47%	47%	69	22%	22%	222	72%	72%	163	53%	53%
F. Moderate	120	39%	86%	190	62%	84%	71	23%	95%	117	38%	91%
F. Low-Very Low	43	14%	100%	49	16%	100%	15	5%	100 %	28	9%	100%
Total	308	100%		308	100%		308	100%		308	100%	

Table 1 shows the frequencies and percentages obtained in the scores of the students who took the CHAEA questionnaire, identifying that the distribution of the data in each of the categories of the scale is varied; few students were found to be located in the low and very low levels, but there is a good accumulation in high and very high. Thus, the sum of the high and very high categories is N=222 of a total of N=308 students, for theoretical learning, which shows us that there is a tendency of the students to be located in this learning style and therefore it is called the predominant learning profile. However, a relevant tendency is observed for pragmatic learning with 163 out of 308 students, followed by active learning with 145 out of 308 students, with a low tendency towards reflective learning with 69 out of 308 students.

Figures 2, 3 and 4 show the results for each style and channel of learning preference according to the preferential tendency, and it is identified that the largest number of students are in high and very high levels of preference in the theoretical and pragmatic channels. This leads us to conclude that the group has a theoretical learning profile, but with a tendency towards mixed learning with a tendency towards the theoretical and pragmatic channels, followed by a moderate tendency towards the active and with a low tendency towards the reflective learning style.

Figure 2.
Preference levels with high and very high tendencies

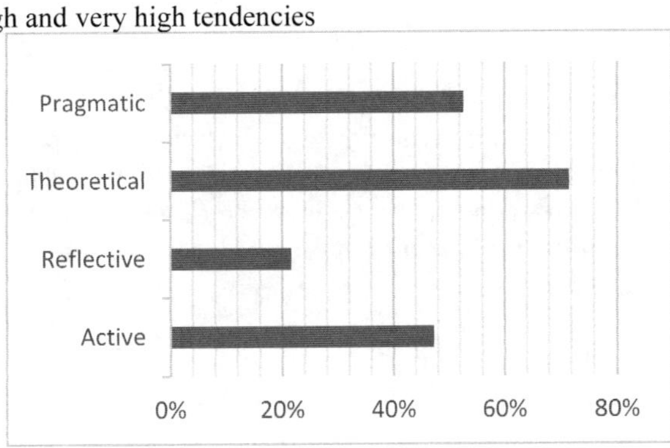

Figure 3.
Preference levels with moderate tendency

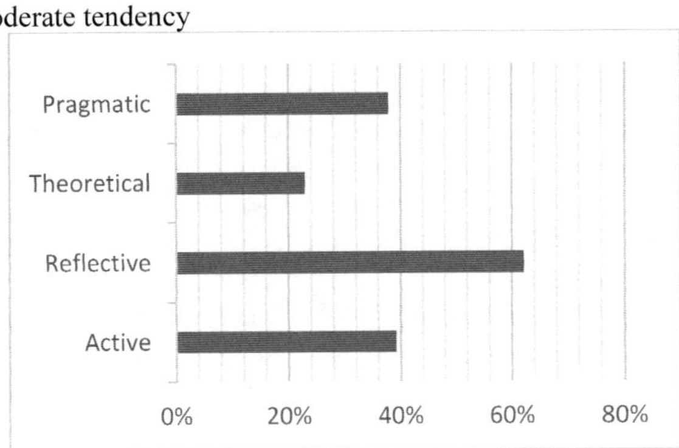

Figure 4.
Preference levels with low and very low tendency

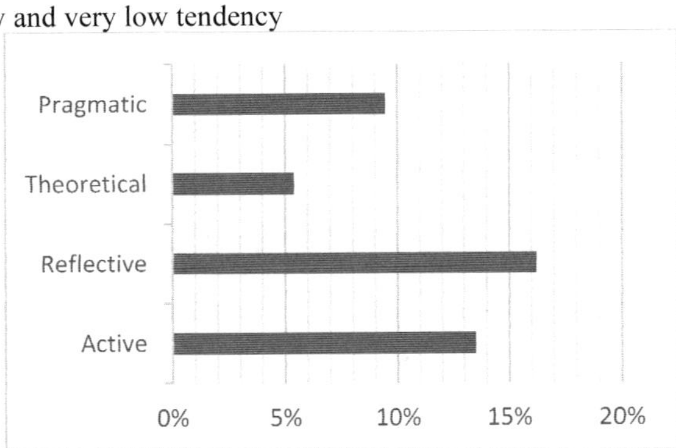

4. Conclusions

The results presented in this research allowed characterizing the learning profile of students taking the course of Computer Science and Technological Convergence of the first semester of systems engineering, virtual modality. In this way, a tendency towards a theoretical learning style can be appreciated, and secondly, a pragmatic-active one. These results coincide with the approaches of González et, al. [6] and Rojas [7], in which it is contemplated that a group of students, especially in the virtual modality, may have a distinctive preference or learning profile, and in turn possess relevant characteristics associated with other styles of preference, given that the individual can learn in a better way through various channels while maintaining a tendency towards the one with the highest preference and use.

The CHAEA questionnaire, by Honey and Alonso [11] (active, reflective, theoretical and pragmatic) based on perception channels, allows determining learning styles and profiles, providing valuable information to determine the predominant learning profile(s), which tends to serve as a basis for the future construction of teaching strategies that may favor better learning results for students, as well as the generation of technological tools that support and optimize this diagnostic process -which would be the next step in this research-.

With respect to the results obtained for this study, it is concluded that the learning style with the highest tendency within the group of N=308 students, is the theoretical (N=222) and followed by the pragmatic (N=163) and the active (N=145). This is then verified in the analysis of relationships between the data of the scales with scores considered in the highest and highest categories, in which the relationship E. A pragmatic and active has a slightly differential value from the relationship that obtained the highest data. These last data, show us that there is diversity of learning styles and perception channels that exist in the group of students, from which it is suggested that the teaching strategies, applied in systems engineering students of the virtual modality should be varied and cannot be focused on a single approach. However, the main tendency towards theoretical learning should be taken into account, while incorporating strategies of the pragmatic and active style; and applying this type of diagnosis as far as possible in the different semesters of the course so that contextualized pedagogical strategies can be applied and possible changes in the tendency over time can be reviewed. This last aspect is perhaps one of the most relevant in the conclusion of this work, which leaves as a proposal, in principle, a methodology to characterize the learning styles and profiles and perception channels of students and opens the way towards the search for a methodology that incorporates technology in a practical way to optimize this type of study so that teachers can perform it without many difficulties, both of analysis and time, which opens the way for future research to measure the learning profile of students and improve teaching processes after making pedagogical strategies based on the information found.

5. References

[1] Sigal, V. (1995). El acceso a la educación superior. Editorial Ministerio de Cultura y Educación, Secretaría de Políticas Universitarias. Buenos Aires (Argentina).

[2] Perrenoud, P. 2004. Diez nuevas competencias para enseñar. Editorial Graó. Barcelona (España).

[3] Dehesa de Gyves, N. (2015). La investigación en el aula en el proceso de formación docente. *Perfiles educativos*, *37*(spe), 17-34. Obtenido de http://www.scielo.org.mx/scielo.php?script=sci_arttext&pid=S0185-

26982015000500003&lng=es&tlng=es.

[4] Imbernón, F. (2012). Nuevos retos del profesorado en el siglo XXI http://www.revista-critica.com/la-revista/monografico/analisis/49-nuevos-retos-del-profesorado-en-el-siglo-xxi

[5] Guaña, E; Llumiquinga, S; Ortiz, K. (2015). Caracterización de entornos virtuales de enseñanza aprendizaje (EVEA) en la educación virtual Ciencias Holguín, XXI, 4, 1-16 Centro de Información y Gestión Tecnológica de Santiago de Cuba Holguín, Cuba

[6] González, K., Padilla, J; Rincón, D. (2012) . Sobre las perspectivas pedagógicas para la educación virtual en Colombia. Obtenido de: https://revistas.pedagogica.edu.co/index.php/TED/article/view/1650

[7] Rojas, J. (2013). *Educación virtual: del discurso teórico a las prácticas pedagógicas en la educación superior colombiana*. e-spacio-UNED . Obtenido de: http://e-spacio.uned.es/fez/view/tesisuned:Educacion-Jerojas

[8] UNESCO. (2014) . Informe de Seguimiento de la EPT en el mundo 2013/4. Enseñanza y aprendizaje. Lograr la calidad para todos. Editorial Organización de la Naciones Unidas para la Educación y la Cultura. Disponible en: http://unesdoc.unesco.org/images/0022/002256/ 225654s.pdf. Consultado 28.08.2015.

[9] UNESCO. (2020). Informe de Seguimiento de la Educación en el Mundo 2020: Inclusión y educación – Todos y todas sin excepción. Organización de la Naciones Unidas para la Educación y la Cultura. Disponible en: https://reliefweb.int/report/world/informe-de-seg[4]

[10] Espinosa, E. (2016). La formación docente en los procesos de mediación didáctica. *Praxis*, *12*(1), 90–102. https://doi.org/10.21676/23897856.1850
uimiento-de-la-educaci-n-en-el-mundo-2020-inclusi-n-y-educaci-n-todos-y

[11] Alonso, C.; Gallego, D. y Honey, P. (1999). Los estilos de aprendizaje. Editorial Mensajero. Bilbao (España).

[12] Giraldo, C. y Bedoya, D. (2006). Los estilos de aprendizaje desde el modelo de V.A.K y su incidencia en el rendimiento académico en niños y niñas de 5to grado de primaria en diferentes estratos socioeconómicos en la ciudad de Pereira de Colombia". Revista Electrónica de Educación y psicología. 2; 4: 1-16. Universidad Tecnológica de Pereira. Pereira (Colombia).

[13] Dunn, R y Dunn, K. (1992). Teaching Elementary Students Through Their Individual Learning Styles. Practical Approaches for Grades 3-6. Massachusetts: Allyn and Bacon (EE.U.U).

[14] Dunn, R y Dunn, K. y Price, G. (1979). Learning Style Inventory (LSI) for Students in grad

[15] Felder, R. y Silverman, L. (2002). "Learning and teaching styles in engineering education" Engineering Education Journal. 78 (7), 674-681. Ed. JEE. International. Disponible en: http://www4. ncsu.edu/unity/lockers/users/f/felder/public/Papers/LS-1988.pdf Consultado 28.08.2015.

[16] Gónzales, B., Alonso, C., y Rangel, R. (2012). El modelo VARK y el diseño de cursos en línea. Revista mexicana de bachillerato a distancia, 4(8), 96 - 103. Recuperado de http://revistas.unam.mx/index.php/rmbd/article/view/44282/40032

[17] Espinoza, J., Miranda, W., y Chafloque, R. (2019). Los estilos de aprendizaje Vark en estudiantes universitarios de las escuelas de negocios. Propósitos y Representaciones, 7(2), 384-414. doi: http://dx.doi.org/10.20511/pyr2019.v7n2.254

[18] Cacheiro, M. (2011). Recursos educativos tic de información, colaboración y aprendizaje. Retrieved 20 February 2021, from https://www.redalyc.org/pdf/368/36818685007.pdf

[19] Zapata, M. (2012). Recursos educativos digitales: conceptos básicos. Recuperado el 6 de octubre de 2015 http://aprendeenlinea.udea.edu.co/boa/contenidos

[20] Garcia, L. (2019): Necesidades de educación digital en un mundo digital. RIED. Revista Iberoamericana de Educación a Distancia.

[21] Dania, C., Marchisio, S. (2013). MODALIDADES DE PERCEPCIÓN SENSORIAL DE ESTUDIANTES DE INGENIERÍA EN SISTEMAS DE INFORMACIÓN. APORTES AL DISEÑO DE MATERIAL DIDÁCTICO PARA LA ENSEÑANZA DE LA ALGORITMIA Invenio. 17 (31-32), 215-228.

[22] Lombardo (2019). Perfiles de aprendizaje. Obtenido de http://integracionneurocorporal.com/aprender-ingles/perfiles-de-aprendizaje/

[23] Valdivia J. (2011). El conocimiento de los estilos de aprendizaje como medida de atención a la diversidad y sus implicaciones educativas en educación infantil. Revista Digital Enfoques Educativos. 75, 85-94.

[24] Ibarra, K. & Eccius, C. (2016). Canales de aprendizaje y su vinculación con los resultados de un examen de ubicación de matemáticas. Revista Intercontinental de Psicología y Educación.16(1), 135-51.

[25] Aguilera, P. (2012). Los estilos de enseñanza, una necesidad para la atención de los estilos de aprendizaje en la educación universitaria, Revista Estilos de Aprendizaje: 10 (1), 79-87).

[26] López, M. (2011). Estilos de aprendizaje, Diferencias por género, curso y titulación, Revista Estilos de Aprendizaje: 7 (7), 109-134

[27] Gutiérrez, M., García, J y Vieira, D. (2012). Estudio de las variables que influyen en los estilos de aprendizaje de diferentes grupos de alumnos de la Universidad de Valladolid, España, Revista Estilos de Aprendizaje: 10 (10), 55-64

[28] Ventura, A.C. Estilos de aprendizaje y prácticas de enseñanza en la universidad. Un binomio que sustenta la calidad educativa, Perfiles Educativos: 33(Esp.) 142-154 (2011).

[29] Bahamón, M., M. Vianchá, L. Alarcón y C. Bohórquez, Estilos y estrategias de aprendizaje relacionadas con el logro académico en estudiantes universitarios, Pensamiento Psicológico: 11(1), 115-129 (2013).

[30] Milgram, R., Dunn, K. y Price, G. (1993). Teaching and Counseling Gifted and Talented Adolescents: An International Learning Style Perspective. 4 (1). 111-182.

[31] Keefe, J. (1988). Aprendiendo Perfiles de Aprendizaje: manual de examinador. Reston, Virginia: NASSP.

[32] Velasco, S. (1996). Preferencias perceptuales de estilo de aprendizaje en cuatro escuelas

primarias. Comparaciones y sugerencias para la formación y actualización de docentes. Revista Mexicana de Investigación Educativa, 1, (2).

[33] Honey, P. y Mumford, A. (1986). The Manual of Learning Styles. Maidenhead, Berkshire: P. Honey, Ardingly House.

[34] McCarthy, B. (1987). The 4MAT system: Teaching to learning styles with right/left mode techniques. Barrington, IL: Excel, Inc.

[35] Fleming, N., y Mills, C. (1992). Not Another Inventory, Rather a Catalyst for Reflection. From To Improve the Academy, 11, 137-155.

[36] Quiñonez, C. (2004). Metodología de estrategia enseñanza-aprendizaje y estilos de aprendizaje. UMBRAL Revista de educación cultura y sociedad. 4 (6), 48-61. Universidad Nacional Pedro Ruiz Gallo y su Facultad de Ciencias Histórico Sociales y Educación. Lambayeque (Perú). Disponible en: http://sisbib.unmsm.edu.pe/bibvirtualdata/publicaciones/umbral/v04_n06/a06.pdf

[37] Hernández, R., Fernández, C. y Baptista, P. (2007), Metodología de la investigación (4ª. ed.). México: McGraw-Hill.

IOP Conf. Series: Materials Science and Engineering 1253 (2022) 012015 doi:10.1088/1757-899X/1253/1/012015

Research methodology to pineapple fiber reinforcement effect in polylactic material

C G Cárdenas-Arias[1,2], E Zuza-Hernández[2], J E Quiroga-Méndez[3], C L Sandoval-Rodriguez[1,5], A D Rincón-Quintero[1,4] and N Y Castillo-León[1]

[1] Design and materials research group (DIMAT), Automation and Control Energy Systems Research Group (GISEAC), Electromechanical engineering, Faculty of Natural Sciences and Engineering, Unidades Tecnológicas de Santander UTS, Calle de los estudiantes No 9-82, 680005, Bucaramanga, Colombia.
[2]Doctoral program in Sustainable Materials and Process Engineering, Department of Sustainable Materials and Process Engineering, University of the Basque Country, UPV/EHU, Engineer Torres Quevedo Plaza, 1, 48013, Bilbao, Spain.
[3]Energy and materials research group, Mechanical engineering school, Universidad Industrial de Santander, Calle 9 Carrera 27, Bucaramanga, Colombia.
[4]Energy in Building research group (ENEDI), Doctoral Program in Energy Efficiency and Sustainability in Engineering and Architecture, Department of Energy Engineering, University of the Basque Country, UPV/EHU, Engineer Torres Quevedo Plaza, 1, 48013, Bilbao, Spain.
[5]Doctoral Program in Electronic and Communications, Department of Communications Engineering, University of the Basque Country, UPV/EHU, 48013, Bilbao, Spain.

ccardenas@correo.uts.edu.co

Abstract. Researching new materials that are environment friendly has become a global purpose. Given the current contamination situation, mainly due to plastic, the using pineapple waste option as a bio composite production source is presented. The methodology used in this project is presented to achieve the material's development, the problems presented and the results obtained in the work progress.

Index Terms: Pineapple, fiber, reinforcement, polylactic material

1. Introduction

Environmental pollution is a problem that is increased mainly by the plastic elements consumption, where the food products commercialization is one that most use them, especially in packaging, which in addition to being non-biodegradable, are harmful to the animal Kingdom health. Additionally, agricultural material such as pineapple is being discarded, which also generates a contamination source, when it's thrown away or when it's burned [1].

It's estimated that 8 million tons of plastic are released into the ocean and that it takes between 450 to more years to break down. In most cases it goes unnoticed because some amphipods eat the bags and turn them into microscopic fragments. These are being ingested by all kind's marine species, from whales to microorganisms, affecting about 700 species, some in extinction danger. Against this background, it's not known whether eating fish is human body dangerous [2].

At the Fourth UN Environment Assembly (UNEA-4) session, held in Nairobi, Kenya, from March 11 to 15, 2019, this point was addressed as one of the event relevant topics. On March 15, 2019, at the United Nations Assembly on the environment, they approved a resolution on plastic garbage and marine microplastics, which is highlighted in point 4 b) [3], "Promote innovation in products design to reduce secondary releases of microplastics from terrestrial and marine sources, and improve waste management when necessary ". This resolution supports the work's intention, as it is to look for a material that can replace plastics with biodegradable materials that minimally affect the environment.

In Colombia, they are interested in abolishing the use of these packages [4], replacing them with cloth or fiber bags, this being a partial solution, since it would be used to load various products from the family basket, and does not point to the accessories industrial development nor the unique packaging products to offer at the community service.

Therefore, it's necessary to supply a material that is friendly with the environment, additionally that does not contaminate food and that has sufficient mechanical characteristics of useful life, resistance, toughness, ductility, to manufacture packages of various sizes and accessories that are being manufactured in plastic [5].

In the development of this composite material, it's planned to use pineapple residues to try to solve another pollution problem that occurs in the areas where this fruit is grown.

2. Methodology

For the work development, four stages are projected, where the finding possibility a biodegradable material based on pineapple will be studied, which can replace plastic as a raw material for products, which are mainly discarded in a very time short period, or that its usefulness is given in a single opportunity.

2.1 Phase 1

The first stage will be dedicated to studying the producing lactic acid feasibility from pineapple residues. To do this, the documentation obtained in the databases will be reviewed, it will be investigated with people who handle the chemical processes concepts and with the fruit growers, with the aim of having a contextualization of what the fruit is, the benefits and problems presented by the waste and learn about the experiences they may have had with the project theme.

Once the result is known, if it's positive, we will find a way to build a pilot plant to produce lactic acid and study the refining of element obtained, the product's characteristics and how to use it in the industry.

2.2 Phase 2

In the second phase it's intended to obtain the polylactic PLA, from lactic acid. Again, the existing bibliography in the different databases should be reviewed and information obtained from primary sources. If the lactic acid from pineapple cannot be used, a lactic acid will be taken that is easily available and that can lead to the investigation.

In this stage, a product or catalyst is sought that allows lactic acid to be converted into a material that has a solid configuration, in order to be able to use it in different industrial applications as a base material.

Once the desired product is achieved, tests will be carried out to characterize it and it will be made available to the people who serve them the properties achieved.

2.3 Phase 3

The third phase is dedicated to obtaining fiber yarns from pineapple leaves. Once again, the bibliography obtained in the different databases should be reviewed and primary information will be obtained in the fruit-producing area.

The different ways of obtaining fiber yarns from different vegetables will be investigated. It is desired to make a shredder machine prototype which allows to obtain the product without mistreating it; the obtaining possibility the thread by hand will be reviewed, taking as a reference the crop incidence area [6].

Once it's obtained, it's estimated to carry out a chemical treatment, which attempts to eliminate the remaining cellulose at the defibration time, which they call whitening, and define by mechanical tests how the material behaves before and after the chemical process [7].

2.4 Phase 4

In the fourth stage, an initial composite material will be integrated, with polylactic matrix with the pineapple threads reinforcement, placing the reinforcement in different directions, in order to mechanically observe the best fiber arrangement with respect to the matrix [8].

To develop the tests and layout define, an experiments design will be specified, reviewing the variables to be taken into account, specifying the standard and specimens' quantities. Once the proper direction of the fiber is determined, the ratio that the fiber should have with respect to the matrix will be reviewed. This result is also obtained by mechanical tests means.

After knowing the fiber proportion with matrix respect, the temperature incidence in the fiber cohesion with the matrix will be studied, being necessary mechanical tests and microscopic observation by SEM.

2.5 Phase 5

Having the material developed, the work's results will be announced, to validate the component with the element's manufacture, preferably a packaging for fruit.

3. Work's advance

In the work's development, some pertinent information has been found regarding the pineapple waste industrial application development, both in obtaining lactic acid, and in the fiber yarns production from the pineapple leaf.

To obtain lactic acid it was found that it is not so simple to elaborate. Very clear and in-depth chemistry, reactions, processes and mixtures knowledge is required.

In 2017 [9], they developed a pilot plant conceptual design to obtain it from sweet products such as sugar cane, but it can be extrapolated due to the pineapple characteristics; In 2019 [10], there is another plant projection, which is described in the document, and in both projections, the plant development is budgeted at more than 100 million COP, which is why its manufacture is not so viable for degree work.

Regarding the fiber production from the pineapple leaves [11], fiber was obtained in a manual artisan way, by beating and combing, where the leaf was tamped with a mallet, it was preliminarily combed to separate the fiber threads, subsequently a water wash was carried out to eliminate cellulose residues and air dried; then with a comb the fibers were obtained.

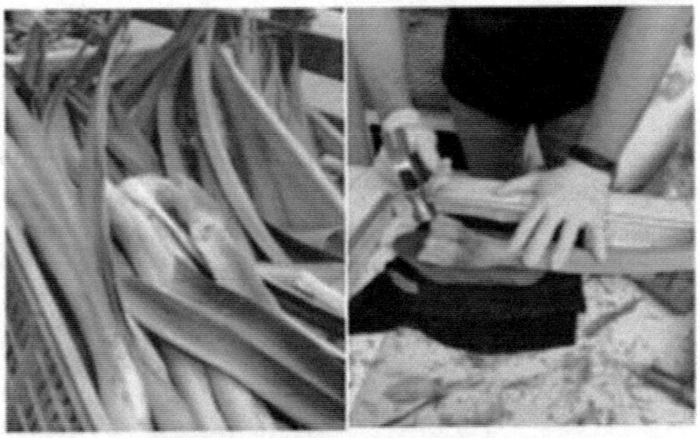

Figure 1. Crushing pineapple leaves

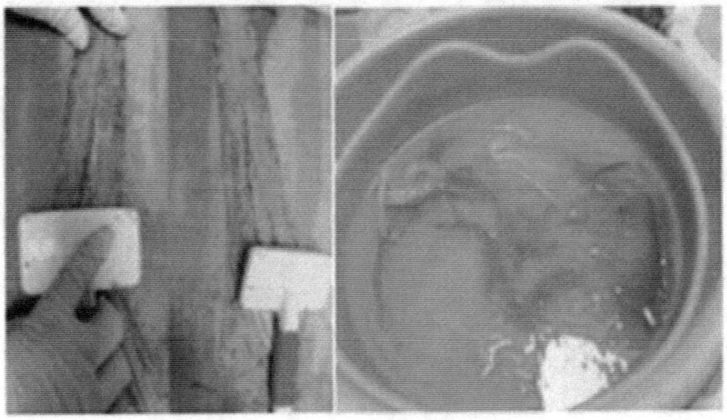

Figure 2. Pineapple fibers cleaning

Figure 3. Fiber hairstyle

A shredding machine was designed, which is under construction, to obtain the fiber in a more industrial way and, incidentally, make a fibers obtained comparison manually with those produced with the machine [12].

A polylactic was generated from industrial lactic acid in liquid presentation, according to the information obtained [13], where the polylactic was generated in an artisanal way, a process during which a polylactic with a gelatinous texture was obtained, which did not comply with the expectations and needs required. Catalysts suggested by chemical engineers were then used, 60% sulfuric acid, stannous chloride, ethanol, and a heating plate and magnetic stirring, where lactide could be obtained, but again when trying to polymerize to generate the PLA, it did not give the required consistency.

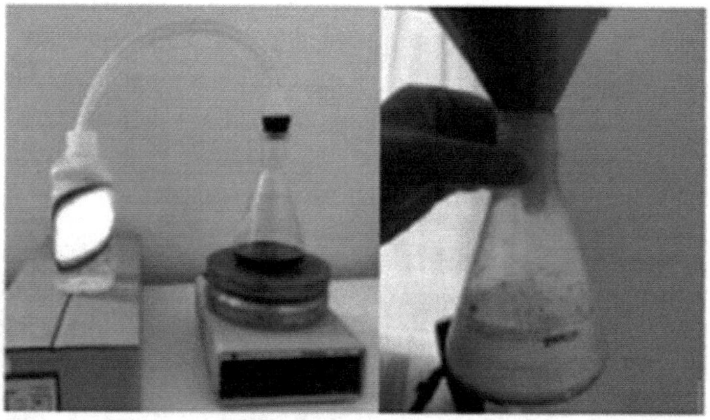

Figure 4. Obtaining lactide

Its then decided to advance in the work, to use industrial PLA in thread form, as the base material.

In the design of experiments, it's defined to produce five specimens with different reinforcement directions, to define the best direction that the fiber should have.

The specimen's measurements are defined, respecting the ASTM D638-14 [14] standard and the NTC 595 standard is revised, which has the same intention.

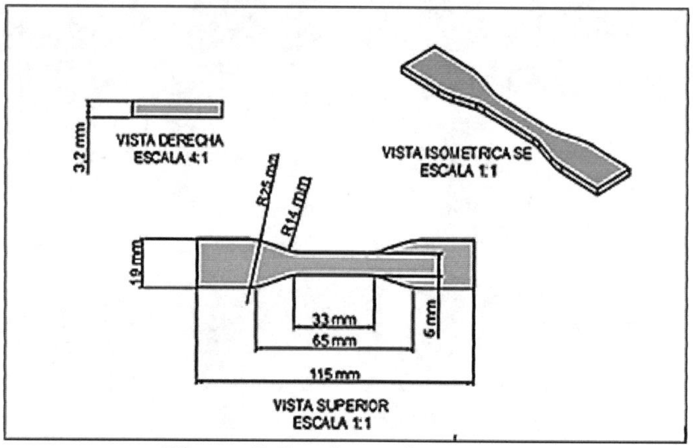

Figure 5. Specimen dimensions

IOP Conf. Series: Materials Science and Engineering 1253 (2022) 012015 doi:10.1088/1757-899X/1253/1/012015

Molds are developed to produce the test tubes based on the NTC 595 [15] standard, in silicone, Teflon and aluminum material.

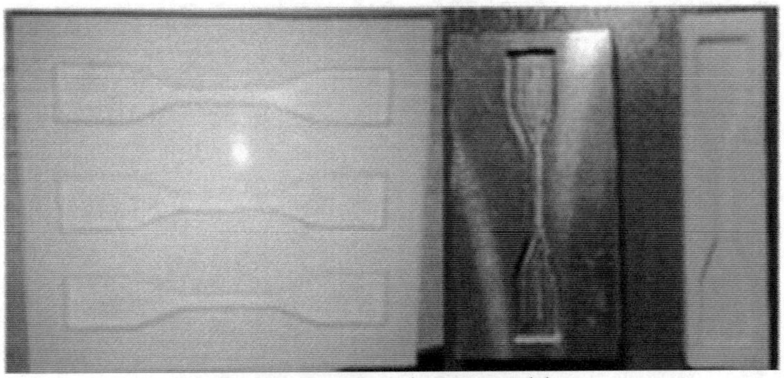

Figure 6. Specimens molds

Five PLA specimens were manufactured without reinforcement, five with longitudinal reinforcement, five with transversely located reinforcement, and five with randomly located reinforcement.

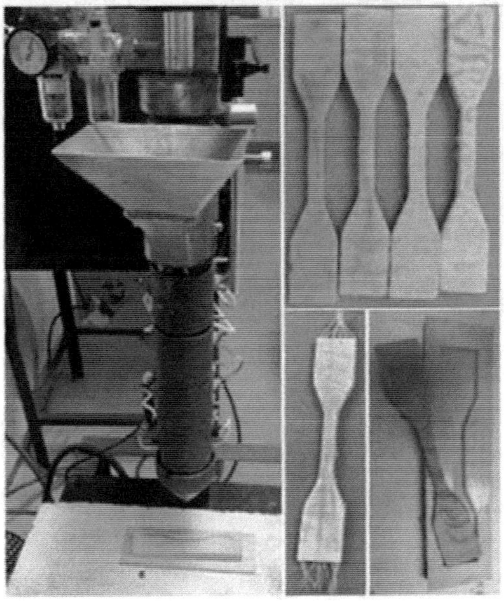

Figure 7. Manufactured test tube

In order to generate the test tubes, injection equipment was used that have resistors along the supply tube, which raise the temperature to 250 ºC to deliver the liquid PLA to the molds. In the mold manually, the fiber was placed in the respective directions according to what was proposed in the experiments design, before pouring the PLA.

The specimens were subjected to tension tests in the UIS laboratories, with a speed of 3 mm / min.

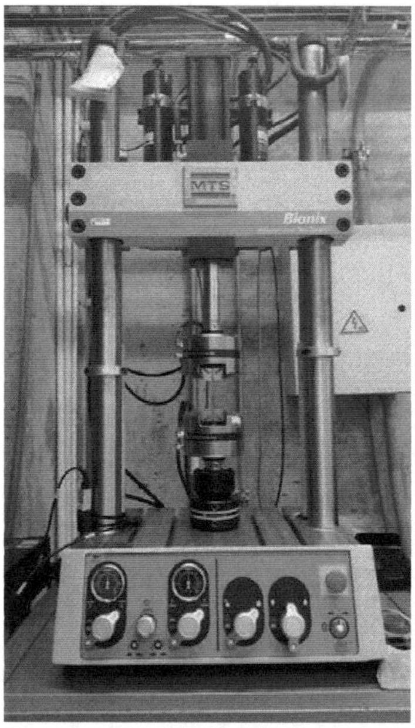

Figure 8. Polymer Tension Machine

As a tests result, it was determined that the best option is the longitudinal fiber location.

Table 1. Comparison table of resistances

Reinforcement type	**Maximum effort** (MPa)
No reinforcement	18.46
Random	16.66
Longitudinal	19.49
Cross	17.52

Once the way to locate the fiber is defined, the best reinforcement matrix ratio will be determined; once this mixture is verified, we will proceed to study the effect that temperature has on the fiber cohesion with the matrix.

4. Conclusions

Whenever a project or work is going to be developed, be it technological development or research, it's necessary to generate a procedure or methodology and there is an expected result. In most cases of this project, what was expected differed with the results that were obtained.

There is a lot of information that is obtained in the networks and databases, which show results, but which is really difficult to repeat, because it's not made explicit how those results were obtained.

It was verified that the best direction of the reinforcing fiber is longitudinal, the other directions decrease the matrix cohesion, generating a decrease in the resistance of the polylactic, while the longitudinal one, contributes the fiber's resistance to that composite material.

The specimen's failure form showed that the fibers avoid the solid elements projection in the breakage event, keeping them together as a security element. Possibly it was due to the cohesion between the fiber and the PLA.

References

[1] O. Salas Murillo, "Pineapple waste: a headache for producers," *Universidad de Costa RIca*, 2018. https://www.ucr.ac.cr/noticias/2018/06/21/desechos-de-la-pina-un-dolor-de-cabeza-para-productores.html.

[2] V. Ingrassia, "Drowned by plastic: pollution in seas and oceans grows alarmingly," *Drowned by Plast. Pollut. seas Ocean. grows Alarm.*, pp. 1–3, 2019, [Online]. Available: https://www.infobae.com/tendencias/2019/08/03/ahogados-por-el-plastico-la-contaminacion-en-mares-y-oceanos-crece-de-forma-alarmante/.

[3] Naciones Unidas, "United Nations Environment Assembly Report of the United Nations Environment Program," pp. 1–5, 2019, [Online]. Available: https://undocs.org/pdf?symbol=es/a/74/25.

[4] C. de Colombia, "Law 1973 of 2019," *El abedul*, vol. 2001, no. agosto 15, pp. 1–109, 2001, [Online]. Available: https://www.minambiente.gov.co/images/normativa/leyes/2001/ley_0685_2001.pdf.

[5] J. J. H. Pariona, "Composites reinforced with natural fibers," *Repos. Univ. Científica del Sur*, 2020, [Online]. Available: https://hdl.handle.net/20.500.12805/1474.

[6] R. J. Lozano Aguillón and S. L. Torres Basto, "Natural fiber production system, based on bibliographical research to take pineapple perolera leaf advantage generated in the Santanderean region," *Repos. Unidades Tecnológicas Santander*, 2021, [Online]. Available: http://repositorio.uts.edu.co:8080/xmlui/handle/123456789/7010.

[7] W. V. Afanador, D. Esteban, and M. Murcia, "Plant distribution in a factory dedicated to obtaining natural thread from the pineapple leaf fiber. (Puerto Rico, Meta).," 2020.

[8] A. Benzerara, M., Guihéneuf, S., Belouettar, R., & Perrot, "Combined and synergic effect of algerian natural fibres and biopolymers on the reinforcement of extruded raw earth," *Constr. Build. Mater.*, 2021.

[9] J. A. Méndez-Alva, E. S. Pérez-Cisneros, and R. Morales-Rodríguez, "Lactic acid production plant conceptual design from sugar industry residues," *Jóvenes En La Cienc.*, vol. 3, no. 2, pp. 736–740, 2017, [Online]. Available: http://www.jovenesenlaciencia.ugto.mx/index.php/jovenesenlaciencia/article/view/2028%0Ahttp://www.jovenesenlaciencia.ugto.mx/index.php/jovenesenlaciencia/article/view/2028/1522.

[10] X. Escobar, "Process bioproduction design of lactic acid from agricultural residues," p. 51, 2019.

[11] C. E. B. Sanchez, "Use of pineapple leaf fibers as reinforcement for biocomposites manufactured by compression molding," UNIVERSIDAD AUTONOMA DE OCCIDENTE, 2017.

[12] E. Of *et al.*, "Artículo original Evaluation of an equipment for the mechanical obtaining of trace fiber from the golden pineapple harvest Resumen," vol. 2, 2020.

[13] N. G. Giaroli and A. A. Maggioni, "Production of Lactic Polyacid by Rop in the Province of Buenos Aires. Study of pre-feasibility," 2015.

[14] Z. Raheem, "Standard Test Method for Tensile Properties of Plastics 1 Standard Test Method for Tensile Properties of Plastics 1," no. January, 2019, doi: 10.1520/D0638-14.

[15] N. T. C. NTC, "NTC 595 Test method to determine the tensile properties of plastics," no. 571, 2007.

IOP Conf. Series: Materials Science and Engineering 1253 (2022) 012016 doi:10.1088/1757-899X/1253/1/012016

Design of a passive optical network test scenario

Alvarado-Jaimes R[1,2], Rincón-Quintero A D[3,4], Sandoval-Rodriguez C L[3,5], Tarazona-Romero B E[3,4] and Ascanio-Villabona J G[3,4]

[1] New technologies group (GNET), Telecomunications engineering, Faculty of Natural Sciences and Engineering, Unidades Tecnológicas de Santander (UTS), Calle de los estudiantes No 9-82, 680005, Bucaramanga, Colombia.

[2] Doctoral Program in Company Internet and Communications Technologies, Universidad las palmas de Gran Canarias, ULPGC, Juan de Quesada, Las Palmas Gran Canarias, España

[3] Design and materials research group (DIMAT), Automation and Control Energy Systems Research Group (GISEAC), Electromechanical engineering, Faculty of Natural Sciences and Engineering, Unidades Tecnológicas de Santander (UTS), Calle de los estudiantes No 9-82, 680005, Bucaramanga, Colombia.

[4] Doctoral Program in Energy Efficiency and Sustainability in Engineering and Architecture, Department of Energy Engineering, University of the Basque Country, UPV/EHU, 48013, Bilbao, Spain.

[5] Doctoral Program in Electronic and Communications, Department of Communications Engineering, University of the Basque Country, UPV/EHU, 48013 Bilbao, Spain.

ralvarado@correo.uts.edu.co

Abstract. Passive optical networks with gigabit ethernet speeds aim to bring internet and communications technologies using fiber optics to the end user. A test scenario for passive optical networks with gigabit ethernet speeds has been implemented. In this test scenario, practices are carried out from the installation of a manager for the administration, operation and maintenance of the active equipment, to the configuration of different types of services over this technology. Unicast and multicast traffic transport tests have been carried out. This test scenario of passive optical networks with low-cost gigabit ethernet speeds allowed simulations with different topologies and network configurations to be carried out and then transferred to real metropolitan networks. The design of the test scenario has been used to simulate the implementation of new technologies on this type of networks before they are implemented in the commercial networks of an Internet service provider. This test scenario has also been used for training in the design and implementation of passive optical networks with gigabit ethernet capabilities.

Index Terms: GPON test scenarios, FTTX networks, Optical communication, Passive optical networks, Gigabit ethernet speeds

1. Introduction.

A passive optical network (PON) is a fiber optic network designed to provide broadband connections to multiple users. Commonly referred to as point-to-multipoint networks, these networks use optical signal splitters from the originating source, called Optical Line Terminals (OLTs), to the end user, known as optical network terminals (ONTs) [1] [2]. PON networks are distinguished from copper networks by their efficiency, cost-effectiveness and applications. For this reason, it is the technology most widely used by Internet Service Providers (ISPs) in the massification of Internet services, voice service and television service [3]. Through this project, PON networks were implemented in the following areas: Internet, voice and television [4].

Through this project, a small-scale GPON network was implemented to perform different laboratory tests of active equipment (OLT - ONT), design simulation and network evaluation when implementing new technologies. Through this lab, Telecommunication companies, ISPs and students in the Telecommunication area can design multiple topologies, perform different network configurations and test new technologies running over GPON networks [5]. The lab design can be used for the implementation and testing of new technologies and services over GPON networks [6].

When this test scenario is used for learning GPON networks, labs are performed for connecting and configuring devices, creating physical and logical topologies, implementing services in real time and ending with best practice exercises and troubleshooting on this type of networks [7].

GPON network simulators allow the design of these networks but are not as reliable in their results when validating results or implementing new technologies on this type of networks. GPON simulators are expensive and do not allow to create practical skills or competencies in the management of this type of networks [8].

Due to the high costs of implementing a GPON network, it was necessary to have a network for testing and validation of services, but through this project it has been demonstrated that it is possible to implement a low-cost PON FTTH test scenario for the development of laboratory practices to serve universities and communications companies [3][9].

2. Work development

Access networks have evolved due to the increase in bandwidth usage by end customers. The use of fiber optic networks and the application of PON (Passive Optical Network) technology in a FTTH (Fiber to the Home) scenario, which integrates multiple services (data, video, voice) in the same fiber strand covering greater distances, is the answer to the need for more bandwidth for Internet users [4][5].

A Passive Optical Network (PON) is a point-to-multipoint optical network (P2MP) called passive, since the components of the distribution network such as optical fiber and signal splitters do not require power source for its operation. PON offers a physical range of 20 km and a division ratio that currently reaches 1:256, i.e., for each PON port it is possible to connect up to 256 users (in the case of XGS-PON technology), it also allows transporting different types of traffic such as ATM, TDM and Ethernet, suitable for home and small business networks [6][7]. Currently, different network topologies called FTTx or Fiber to the X can be implemented, depending on the reach of the optical fiber. Thus, we can find different types of FTTx networks such as:

FTTC - C: Curb. Corner or Cabinet
FTTH - H: Home. Home
FTTO - O: Office. Office
FTTM - M: Mobile. Mobile Networks

The most widely known and deployed technology in commercial networks is FTTH GPON, a Gigabit-capable PON network that reaches up to a communications device located in the home and is defined by the International Telecommunication Union's ITU-T G.984.X series.

There are different types of software available for GPON network design. Unfortunately, this software is not useful for training technicians or engineers to implement this type of network. Another deficiency of the software is the implementation of new services over GPON networks, since they are not supported. In these cases, there is a need to have a test scenario that serves to train in practical skills, validate designs in this type of network and implement new technologies that work on GPON technology.

The experimental design of the practice scenario of this project includes the installation of the EMS Element Management System for the administration, operation and maintenance of the OLT and ONTs, and the configuration of different test scenarios including traffic without VLAN, traffic with VLAN, Multicast traffic and the configuration of an Internet access service. The most relevant laboratory practices that can be performed in the designed scenario are described in Table 1.

Table 1. Simulations and test scenarios

Practices	Description
1	EMS installation
2	Physical connection to the OLT
3	ONT Detection and Aggregation
4	Transporting traffic without VLAN
5	Transporting traffic with VLAN
6	Fixed IP Internet Service Configuration
7	Multicast service configuration

Note: Table prepared by the authors of the project.

Each of the test scenarios was developed taking into account the designs and configurations of telecommunication companies that offer GPON technology services. These scenarios can be used for teaching the implementation and configuration of GPON networks. They are also scenarios where new technologies can be validated over fiber optic networks. This allows companies to evaluate their future implementation in the networks serving users without affecting availability.[10]

IOP Conf. Series: Materials Science and Engineering 1253 (2022) 012016 doi:10.1088/1757-899X/1253/1/012016

Many companies and universities have wanted to implement a laboratory and scenarios of GPON networks either for their operators or students to practice in their design and installation. Also, a lab of this type is useful for study, analysis and validation with new services. Unfortunately, the prices of emerging technologies are expensive and so is this type of lab. A study of the market for GPON network equipment and devices has been carried out and contrasted with the needs of the company and the university in order to acquire the ideal elements.[11]

For the PON test scenario, the telecommunications laboratory has the minimum necessary elements described in Table 2.

Table 2. Inventory of equipment delivered to the Laboratory.

Quantity	Description
1	OLT EPON 2 x PONs, 2 x electrical GE and 2 x optical GEs
2	SFP EPON
4	ONT xPON de 2 y 4 Gb Ethernet port Wi-Fi
2	*Splitter* 1:8
2	Attenuator SC/PC
6	Patch-cord SC/PC-SC/APC

Note: Table elaborated by the authors based on the elements acquired for delivery to the Telecommunications Laboratory of the Unidades Tecnológicas de Santander.

3. Internship procedure of laboratory

The proposed scenario allows developing practical skills for the installation of optical fiber and other components in a GPON network. This scenario also allows the implementation of different topology designs and device configurations in GPON networks. In addition, it allows the testing of new services and applications that work over GPON networks.

In the first stage of the practices with the GPON laboratory, the installation and management of the Element Management System (EMS) software from a local server is performed. In the second stage, practice is given in the management of the physical connections through console and network cable for the configuration of the OLT. The EMS server allows professionals to manage the FD1002S OLT remotely using an EMS client to access the server. The third stage involves the detection and aggregation of ONTs. The fourth stage involves the configuration of virtual local area networks (VLANs) for traffic transport as shown in Figure 1.

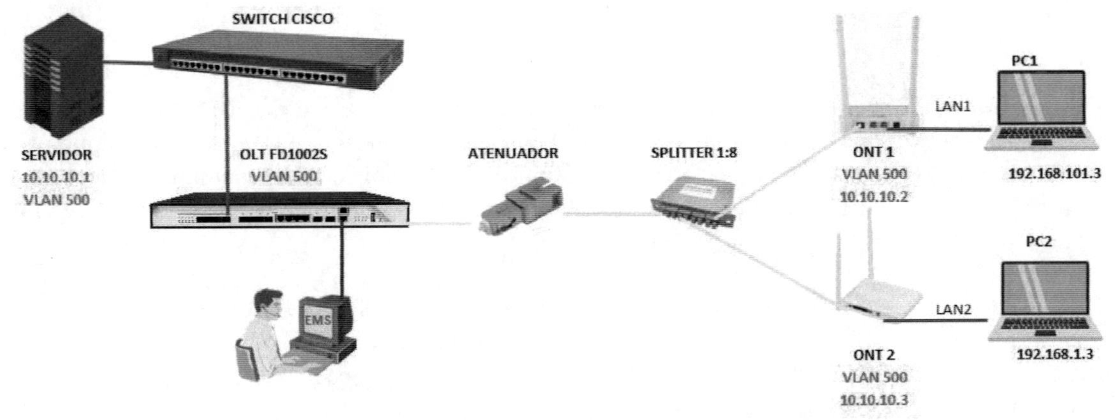

Figure 1. Topology configuration with local server for GPON

In the last practices the connection to the Internet is implemented using static routes or routing protocols and the sixth and final stage is for traffic management to provide television service through Multicast broadcasting using the topology with Internet connection in Figure 2.

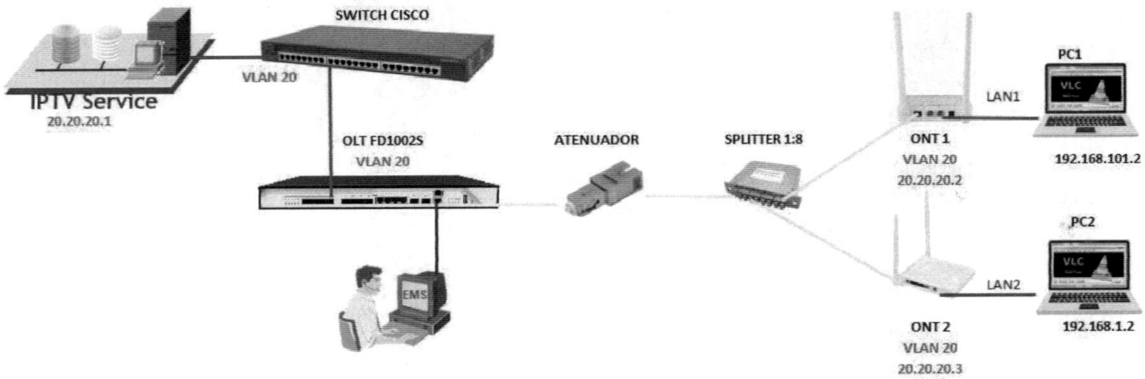

Figure 2. Configuration with multicast service via Internet connection

After performing the different sequential practices for the learning and use of GPON equipment, a manual of good practices and troubleshooting is made. For this and on the topology with a video server equipment connected to the OLT and a video client connected to an ONT, the system is verified using the intermediary equipment and the EMS.

4. Conclusions

This scenario of GPON FTTH networks has been used to test and validate the implementation of services over this type of networks. This scenario has been used to analyze the design of GPON network topologies. This GPON network scenario is low cost and is an option to be implemented in companies and higher education institutions for learning in the design and implementation, configuration and maintenance of this type of networks.

The didactic test bench achieved the purpose of the work by establishing a real low-cost scenario for practical testing of an FTTH/PON network, allowing interaction with active and passive elements. This project has provided the possibility to explore aspects of the network management area by means of EMS software. The proposed didactic testbed provides a clear path of activities and procedures associated with the implementation, operation and maintenance of FTTH/PON networks.

References

[1] Z. Abdellaoui, Y. Dieudonne, y A. Aleya, «Design, implementation and evaluation of a Fiber To The Home (FTTH) access network based on a Giga Passive Optical Network GPON», *Array*, vol. 10, p. 100058, jul. 2021, doi: 10.1016/j.array.2021.100058.

[2] I. Ogushi, N. Araki, y Y. Azuma, «A New Optical Fiber Line Testing Function for Service Construction Work That Checks for Optical Filters Below an Optical Splitter in a PON», *J. Lightwave Technol.*, vol. 27, n.º 23, pp. 5385-5393, dic. 2009, doi: 10.1109/JLT.2009.2032305.

[3] S. Hoernig, S. Jay, K.-H. Neumann, M. Peitz, T. Plückebaum, y I. Vogelsang, «The impact of different fibre access network technologies on cost, competition and welfare», *Telecommunications Policy*, vol. 36, n.º 2, pp. 96-112, mar. 2012, doi: 10.1016/j.telpol.2011.12.003.

[4] K. S. Kim, K. Ennser, y Y. K. Dwivedi, «Clean-slate design of next-generation optical access», en *2011 13th International Conference on Transparent Optical Networks*, Stockholm, Sweden, jun. 2011, pp. 1-4. doi: 10.1109/ICTON.2011.5970910.

[5] K. A. Mat Sharif, N. A Ngah, A. Ahmad, K. Khairi, Z. A. Manaf, y D. Tarsono, «Demonstration of XGS-PON and GPON Co-Existing in the Same Passive Optical Network», en *2018 IEEE 7th International Conference on Photonics (ICP)*, Kuah, abr. 2018, pp. 1-3. doi: 10.1109/ICP.2018.8533167.

[6] H. A. Bakarman, S. Shaari, y M. Ismail, «Simulation of 1.25 Gb/s downstream transmission performance of GPON-FTTx», en *International Conference On Photonics 2010*, Langkawi, Malaysia, jul. 2010, pp. 1-5. doi: 10.1109/ICP.2010.5604440.

[7] N. Merayo *et al.*, «Experimental validation of an SDN residential network management proposal over a GPON testbed», *Optical Switching and Networking*, vol. 42, p. 100631, nov. 2021, doi: 10.1016/j.osn.2021.100631.

[8] L. Malina, T. Horvath, P. Munster, y J. Hajny, «Security solution with signal propagation measurement for Gigabit Passive Optical Networks», *Optik*, vol. 127, n.º 16, pp. 6715-6725, ago. 2016, doi: 10.1016/j.ijleo.2016.04.069.

[9] J. Prat *et al.*, «Demonstration and field trial of a resilient hybrid NG-PON test-bed», *Optical Fiber Technology*, vol. 20, n.º 5, pp. 537-546, oct. 2014, doi: 10.1016/j.yofte.2014.06.006.

[10] Y. Yi *et al.*, «Simulations and Experiments on the Effective Optical Gain of FEC in a GPON Uplink», *IEEE Photon. Technol. Lett.*, vol. 19, n.º 1, pp. 82-84, ene. 2007, doi: 10.1109/LPT.2006.888969.

[11] A. Israr, M. Junaid, y A. Israr, «Performance Analysis of Advance Optical Modulation Formats for GPON System», en *2015 13th International Conference on Frontiers of Information Technology (FIT)*, Islamabad, Pakistan, dic. 2015, pp. 77-80. doi: 10.1109/FIT.2015.11.

Evaluation of control strategies applied in small-scale photovoltaic solar tracking systems: a review

B E Tarazona-Romero[1 2], E J Plata-Pineda [1], C L Sandoval-Rodriguez[1 3], J A Ascanio-Villabona[1 2], and O Lengerke-Péreza [1]

[1] Automation and Control Energy Systems Research Group (GISEAC), Faculty of Natural Sciences and Engineering, Electromechanical engineering, Unidades Tecnologicas de Santander (UTS), Student Street No 9-82, 680005, Bucaramanga, Colombia.
[2] Energy in Building research group (ENEDI), Doctoral Program in Energy Efficiency and Sustainability in Engineering and Architecture, Department of Machines and Thermal Engines, University of the Basque Country (UPV / EHU), Engineer Torres Quevedo Square, 1, 48013, Bilbao, BI, Spain.
[3] Doctoral Program in Electronic and Communications, Department of Communications Engineering, University of the Basque Country, UPV/EHU, 48013 Bilbao, Spain.

Abstract. This work evaluates the control algorithms applied to decentralized photovoltaic solar tracking systems. For this, the control strategies are divided into three: open loop, closed loop and hybrid loop. Each strategy is screened for its suitability as a suitable technology for small-scale applications, defined by meeting a set of eleven attributes. A specialized literature review process was developed specifically to present the potential of each control algorithm strategy, through an evaluation matrix. The analysis of the algorithms and attributes was carried out using the formal methodology of concept analysis. To facilitate the processing of the information, free access software called "concept Explorer" is used. The analysis carried out shows that open loop control algorithms currently have a greater application in one and two-axis solar tracking systems. Additionally, the on-off control is the one with the greatest application for the three types of control loops evaluated. Finally, the applied methodology has proven to be useful for the evaluation of information and serves as a reference, to carry out analyzes that group and link different alternatives as a model for evaluating a deterministic set of attributes.

Keywords: Control algorithm, Solar Trackers, Formal Analysis of Concepts, Photovoltaic Systems, Renewable Energy

1. Introduction

In recent years, solar photovoltaic (PV) capture systems have presented great technological advances[1] [2]. These systems are generally characterized by being systems that capture solar radiation directly "statically" and convert it into electricity; considered as conventional. In order to improve the efficiency of static PV systems, which are directly affected by solar variation, studies have been carried out with PV systems that follow the solar path, significantly improving the efficiency of the systems[3].

Solar tracking systems (ST) can be classified according to their degrees of freedom, in single-axis or multi-axis systems [4] [5]. The application of these systems can significantly increase the amount of energy produced, between 12% and 25% for PV systems with single axis solar tracking (ST) and between 30% and 40% for a solar tracking system. with two axles[5].

However, ST´s still have limitations, which only ST can overcome if optimized and precisely controlled. The main focus to improve the control in the trackers, should be put in the configuration of the tracking axes, optimizing the mobile devices and the appropriate configuration of the control systems [6]. In recent years, specialized computer technologies and reliable control systems have opened the window to generate studies with the aim of improving the design and operational algorithms of solar tracking systems (ST), in order to improve efficiency of photovoltaic systems [7].

Consequently, some works have already analyzed different aspects focused on solar tracking systems. In the work of Khalil et al., a review of solar monitoring techniques for PV systems was presented, evaluating their efficiency, as well as advantages and disadvantages of control methods [8]. In another review, Sumathi et al., analyzed ST methods and solar tracking algorithms, concluding that systems with two-axis solar tracking have an efficiency of 30% higher than static PV systems [9].

Subsequently, Motahhir et al., presented a detailed review of the MPPT algorithms for ST where they evaluate their characteristics, suitability and application[10]. In turn, other authors analyzed and classified ST technologies, discussing advantages and disadvantages of their application [11] [12] [13].

On the other hand, recently, the International Electrotechnical Commissions (IEC) created the IEC 62817 standard, which is an international standard to qualify the design of solar trackers under a standardized methodology [14]. Some authors applied the standard in their technological developments and research, identifying some problems and proposed a series of improvements for future updates of the standard [15] [16] [17].

Eventually, this work presents a theoretical analysis of the current state of control algorithms applied to low-scale solar tracking systems. The evaluation of the information will be developed under the application of the formal concept analysis methodology (FCA), through the use of the concept Explorer software, relating attributes and objects, facilitating the treatment of information and its analysis.

Finally, section 2 evidences the methodology applied to the review, which was based on the FCA information classification method. For its part, section 3 will present the classification of the different control algorithms applied to solar monitoring systems and their compliance or not, with each of the attributes raised in section 4. Subsequently, section 4 presents the results of the work. . Finally, section 5 presents the most relevant conclusions of the work.

2. Review Methodology

The information evaluation process is carried out based on the formal analysis methodology of Concepts, which allows relating a series of attributes that each of the objects to be studied must meet. This relationship makes it possible to determine which object fulfills more attributes, showing, in turn, the attributes shared by the objects.

For the evaluation presented in this document, the following types of algorithms, classified as follows, will be related as objects:

i. Open loop
ii. Closed loop
iii. Hybrid loop (closed loop and open loop)

Eventually, the objects will be classified into the following attributes:

A1. Single Axis Tracking System
A2. Two-axis Tracking System
A3.Classic control algorithm

- On-Off
- PI (Proportional-Integral)
- PID (Proportional-Integral-Derivative)

- SMC (Slider Mode Control)

A4. Modern control algorithm
- PID-FLC (Proportional-Integral-Derivative + Fuzzy Logic Control)
- Fuzzy Logic Control (FLC)
- Light dependent resistors (LDR)
- Neural Network
- Adaptive Sliding Mode Control (SMC)
- IMC-PID (Internal + Proportional-Integral-Derivative Model Control)
- Neuro-fuzzy logic
- Sliding mode
- Parametric optimization
- Prophetic
- Linear quadratic regulator
- Machine Learning
- Robust Control
- Bayesian network
- Logic-based supervisor
- Waterfall

A5. Basic control unit
- Arduino
- Basic electronic
- Microcontroller
- Raspberry Pi
- Smart relay
- PLC (Programmable Logic Controller)

A6. Advanced control unit
- Specialized software application "Matlab", "Simulink", "LabVIEW"
- Robust PLC

A7. Availability of materials locally
A8. Open access software application
A9. Low investment cost
A10. Low impact on Tracking Error
A11. High application in photovoltaic systems

Considering objects and attributes, each object is evaluated by each of the attributes listed above. This will give rise to a matrix that will allow feeding the concept Explorer tool, allowing a qualitative evaluation of the information.

3. Clasificación de los algoritmos de seguimiento solar

The types of algorithms followed to be analyzed were evaluated individually in order to respond to each of the attributes raised and thus finally define whether or not they meet this condition. This section will be divided into three: open-loop algorithm, closed-loop algorithm, and hybrid algorithm.

3.1. Open Loop Algorithm
This solar tracking strategy is called open-loop because there is no control feedback through the application of solar sensors. Figure 1 shows the block diagram of an open loop strategy for solar trackers.

Figure 1. Block Diagram Open Loop Solar Tracking Strategy

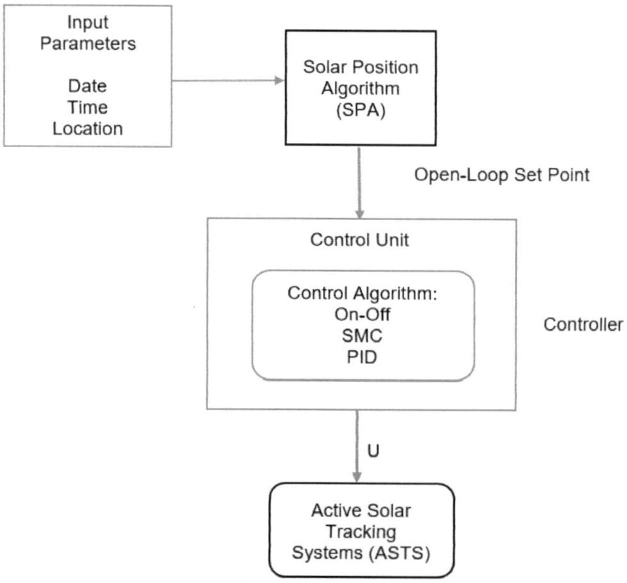

Source: Author

Generally, open loop solar tracking systems operate under the strategy of Solar Positioning Algorithms (SPA), based on solar coordinates that predict the apparent position of the sun with respect to a geographic location of the system on Earth for one-axis solar trackers or two axles [18] [3]. These algorithms are conventionally programmed in electronic control units through open access tools and the use of Hardware with high availability at local levels such as: Arduino, microcontrollers, programmable logic controller (PLC), smart relay, Raspberry Pi, among others [19] [3].

Once the algorithm programming is developed in the software and linked to the hardware, the algorithm calculates the solar zenith and azimuth angles that allow predicting the solar position through a control signal (U), which align the solar panel photovoltaic or thermo solar system to solar vector. It is important to mention that they are highly applied to small-scale photovoltaic systems [3].

On the other hand, Chong et al., Frathabadi et al., Pisirir and Bingol, calculated the closed loop solar tracking error ($STE_{OpenLoop}$) using equation 1 [20] [21] [22], which facilitates the establishment of the frequency of the tracking systems or the hysteresis of monitoring that allows optimizing energy production and consumption in the control algorithms applied to solar monitoring systems [23]:

$$STE_{OpenLoop} = SP_{Tracker} - SP_{SPA} \quad (1)$$

The open-loop strategy is of low precision in solar tracking compared to the closed-loop and hybrid strategy, this is due to the instability of the system when faced with an electrical or mechanical disturbance[24] [21] [25]. Additionally, it presents inaccuracies in the solar tracking algorithm, due to the input parameters that are: geographic location, time, date and orientation [26] [3] [19]. To this is added, errors in the process of transforming the solar coordinates into rotational angles of the actuators and tolerances of the manufacturing, assembly and installation processes of the elements that make up the system [21] [27] [28].

The implementation of the open loop algorithm is easy and inexpensive, it does not require the use of a solar sensor and its operation is independent of meteorological alterations, avoiding the error introduced by the use of photosensors in closed circuit systems [9] [3]. Additionally, some authors included a GPS as an alternative, to improve the input data of the geographic parameters to the algorithm, reducing this error, but increasing the costs of the system [29] [30].

Next, the relevant works that link in their experimental or simulation process, open loop control algorithms applied to solar tracking systems are presented (Ver Table 1). Es indispensable destacar, que no todos los autores incluyen en sus trabajos el porcentaje (%) de error del sistema de seguimiento solar.

Table 1. Experimentation and simulations with open loop control algorithms applied to solar trackers

Item	Author	Exp. / Sim.	Control algorithm	Axis	Solar location	Precision
1	[17]	Experimental	On-Off	One	Solar control algorithm	+/- 0.4°
2	[21]	Experimental	On-Off	Two	Solar control algorithm	+/- 0.43°
3	[27]	Experimental	On-Off	Two	Solar control algorithm	+/- 0.43°
4	[31]	Experimental	On-Off	Two	Solar control algorithm	>0.5 °
5	[22]	Experimental	On-Off	Two	Solar control algorithm	/
6	[6]	Experimental	On-Off	Two	Solar control algorithm	>0.2 °
7	[32]	Simulation	SMC	Two	Solar control algorithm	/
8	[30]	Experimental	PID	Two	Solar control algorithm	+/- 0.5°
9	[29]	Experimental	On-Off	Two	Solar control algorithm + GPS	/
10	[33]	Simulation	PID	Two	Solar control algorithm	/
11	[34]	Experimental	On-Off	One	Solar control algorithm	/
12	[28]	Simulation	PID	Two	Solar control algorithm	+/- 0.17° / +/- 0.56°
13	[35]	Experimental	On-Off	Two	Solar control algorithm	/
14	[36]	Experimental	On-Off	Two	Solar control algorithm	/
15	[37]	Experimental	On-Off	Two	Solar control algorithm	+/- 0.6°
16	[18]	Simulation	Kinematic synthesis	Two	Solar control algorithm	/
17	[38]	Experimental	FLC	Two	Solar control algorithm	/
18	[25]	Simulation	ANFIS	One-Two	Solar control algorithm	/
19	[39]	Experimental	On-Off	Two	Solar control algorithm	+/- 0.2°
20	[40]	Experimental	On-Off	Two	Solar control algorithm	

Source: Author

3.2. *Closed loop algorithm*

This solar tracking strategy is called closed-loop because there is control feedback of the apparent position of the sun through the application of solar sensors and it is applied in one-axis or two-axis solar trackers [41]. Figure 2 shows the block diagram of a conventional closed-loop strategy for solar trackers.

Generally, closed-loop solar tracking systems depend on the solar sensor and require setting a set point, typically equal to 0, based on arrays of photosensors or the center of an image (0,0) for solar imaging Sensors [41]

[42] [3]. Salgado- in his work, highlights that most closed circuits use a photosensors array and that it is possible to calculate the closed-loop solar tracking error ($STE_{ClosedLoop}$), with the set point established and the position of the sun measured by the solar sensor (SP_{sensor}), using equation 2 [43]:

$$STE_{ClosedLoop} = 0 - SP_{sensor} \qquad (2)$$

Figure 2. Block diagram of a conventional closed-loop solar tracking strategy

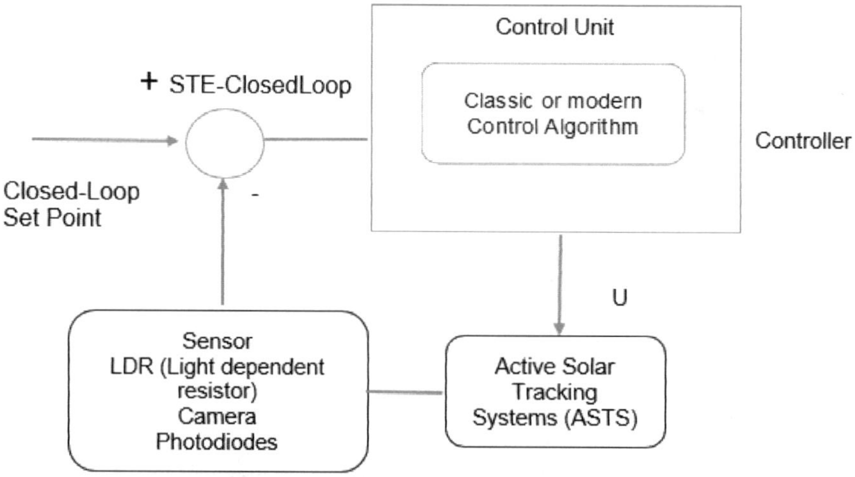

Source: Author

These algorithms, after the process of determining the solar tracking error, feed the algorithm to determine the solar zenith angle and azimuth to locate the solar position [44]. This algorithm is programmed conventionally in electronic control units through open access tools and the use of Hardware with high availability at local levels such as: Arduino, microcontrollers, programmable logic controller (PLC), smart relay, Raspberry Pi, among others [8] [44] [3], as well as with modern control algorithms for special cases [3].

The closed-loop strategy has a great advantage and is that it achieves greater precision than open-loop systems due to feedback control by integrating the solar sensor into the system [45] [46]. Additionally, they do not require an algorithm for solar tracking or prediction of the apparent position of the sun and they do not restrict geographic location [47] [48]. Its application has been well received in concentrating solar and photovoltaic systems [3].

However, it requires a solar sensor, which can be affected by meteorological disturbances and diffuse irradiance, so it can be said that its operation depends on meteorological alterations [38]. The solar location is limited to the field vision of the sensor and most of the sensors applied in this technique do not have a linear response to different temperature and radiation levels [41] [49].

Additionally, the errors in the process of transforming the solar coordinates into rotational angles of the actuators and tolerances of the manufacturing, assembly and installation processes of the elements that make up the system [50]. As for the implementation of the open-loop algorithm, it is considered complex and the cost of implementation is higher than that of an open-loop system [51].

Next, the relevant works that link in their experimental or simulation process, closed-loop control algorithms applied to solar tracking systems are presented (See Table 2). It is essential to note that not all authors include the error percentage (%) of the solar tracking system in their work.

Table 2. Experimentation and simulation with closed loop control algorithms applied to solar trackers

Item	Author	Exp. / Sim.	Control algorithm	Axis	Solar location	Precision
1	[52]	Experimental	On-Off	Two	LDR	/
2	[53]	Experimental	On-Off	Two	LDR	/
3	[45]	Simulation	LQR	Two	/	/
4	[6]	Experimental	On-Off	Two	LDR	> 2°
5	[54]	Experimental	On-Off	Two	Pi Camera Lens	/

6	[21]	Experimental	On-Off	Two	LDR	0.14°
7	[55]	Simulación	On-Off	Two	/	/
8	[56]	Experimental	PID	Two	/	/
9	[57]	Simulation	SMC	Two	Photosensors	/
10	[58]	Simulation	PID/LFC	One	LDR	/
11	[59]	Experimental	Neuro Fuzzy Logic	Two	Photodetector	/
12	[50]	Experimental	On-Off	Two	Photosensors	/
13	[60]	Experimental	On-Off	Two	Irradiance sensor	/
14	[61]	Experimental	On-Off	Two	LDR	/
15	[62]	Experimental	PI	Two	Matriz de Photodiodes	/
16	[63]	Experimental	On-Off	Two	LDR	/
17	[51]	Experimental	On-Off	Two	LDR	/
18	[64]	Experimental	On-Off	One	LDR	/
19	[65]	Experimental	On-Off	Two	Photosensors	/
20	[41]	Experimental	On-Off	Two	Webcam	/
21	[48]	Experimental	PI	Two	LDR	/
22	[37]	Experimental	On-Off	Two	Camera	/
23	[66]	Experimental	FLC	Two	LDR	0.91 ° / 0.27°
24	[46]	Simulation	SMC	Two	/	/
25	[10]	Experimental	On-Off	Two	LDR	/
26	[49]	Experimental	On-Off	Two	LDR	/
27	[67]	Experimental	On-Off	One	LDR	1.5 °
28	[68]	Experimental	On-Off	One	LDR	/
29	[38]	Experimental	FLC	Two	LDR	0.10 ° / 0.45°
30	[69]	Experimental Simulation	PID/LFC	Two	Solar control algorithm – Gyroscope	/
31	[70]	Experimental	On-Off	One	UV sensor	/
32	[71]	Experimental	On-Off	Two	LDR – UV sensor	/
33	[72]	Experimental	PI	One	LDR	

Source: Author

3.3. Hybrid Algorithm

This solar tracking strategy is called hybrid loop because it integrates closed loop and open loop strategies. It arose based on the requirements demanded by solar concentration systems, but it is applied in one-axis and two-axis photovoltaic systems. Figure 3 shows the block diagram of a conventional hybrid loop strategy for solar trackers. Typically, this strategy is made up of two control loops: a robust control implemented in an open loop and a fine control implemented in a closed loop.

Generally, the open loop and the closed loop integrated in this technique are not activated at the same time, however some authors use weighting algorithms between loops to determine the apparent position of the sun [73] and others use different algorithms to carry out the analysis. solar tracking [74].

Figure 3 shows that there are two possibilities to establish a set point or "set point" of the control system; open loop and closed loop. For this strategy, the control unit is able to change modes in real time depending on the location stage of the solar tracker. Subsequently, the control unit sends a signal to the actuators in order to generate an action, which positions the system with respect to the solar position.

This algorithm is programmed conventionally in robust and classic electronic control units, integrating the two tools through open and closed access software [75] [76]. The use of Hardware presents availability at local levels in some pieces such as: Arduino, microcontrollers, programmable logic controller (PLC), smart relay, Raspberry Pi, among others [8] [44] [3], as well as import requirements by purchase of robust equipment and specialized software [3].

Figure 3. Block diagram of a conventional hybrid-loop solar tracking strategy

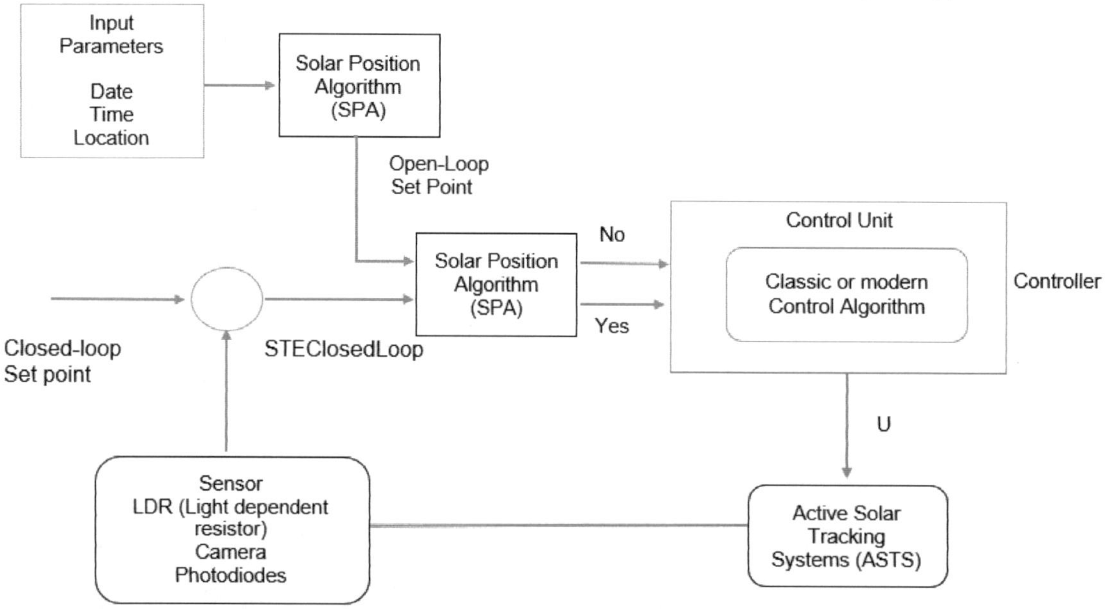

Source: Author

The main advantage of the hybrid loop is that it is not affected by factors that decrease the accuracy of the solar tracking system, such as the installation deviation of the tracker, errors in the estimation of the position of the sun by the open loop control algorithm. and errors due to weather conditions [77] [78]. Consequently, it is a strategy with a higher level of reliability and precision compared to open or closed loop [79].

The implementation is highly expensive compared to the closed or open loop strategy, as well as its complexity at the time of assembly and its operation [80] [81]. Additionally, it requires a solar tracking algorithm and a solar sensor. Due to the cost of implementation, only centralized solar concentrating systems are widely used, with few applications in photovoltaic systems [82] [83].

Next, the relevant works that link in their experimental or simulation process, hybrid loop control algorithms applied to solar tracking systems are presented (See Table 3). It is essential to note that not all authors include the error percentage (%) of the solar tracking system in their work.

Table 3. Experimentation and simulation with hybrid loop control algorithms applied to solar trackers

Item	Autor	Exp. / Sim.	Control algorithm	Axis	Solar location	Precision
1	[84]					
2	[85]	Experimental	On-Off	Two	Solar control algorithm and Photodiode	<0.3°
	[77]	Experimental	On-Off	One/Two	Solar control algorithm and Photodiode	/
3	[80]	Experimental	On-Off	Two	Solar control algorithm and LDR	+/- 2°
4	[75]	Simulation	On-Off	Two	Solar control algorithm an solar sensor	<0.91°
5	[73]	Experimental	PID	Two	Solar control algorithm and LDR	+/- 0.2°

6	[81]	Experimental	PID	Two	Solar control algorithm and Photodiode	<0.15°
7	[82]	Simulation	PID	Two	Algoritmo de control solar and CCD	/
8	[76]	Experimental	On-Off	Two	Solar control algorithm and Climate sensor	/
9	[78]	Experimental	On-Off	Two	Solar control algorithm with GPS and Photosensors	<0.91°
10	[83]	Simulation	Bayesian Network	/	Solar control algorithm	/
11	[86]	Simulation	On-Off	Two	Solar control algorithm	/
12	[87]	Simulation	On-Off	Two	Solar control algorithm	/
13	[64]	Simulation	On-Off	One	Solar control algorithm	/

Source: Author

Finally, Table 4 presents the attribute matrix for the three control algorithm techniques applied to solar trackers. The attributes that meet for each technique are identified with an X within the matrix.

Table 4. Selection matrix of attributes with respect to objects

At / Ob	A1	A2	A3	A4	A5	A6	A7	A8	A9	A10	A11
Open Loop	X	X	X		X		X	X	X		X
Closed Loop	X	X	X	X	X	X	X	X		X	X
Hybrid Loop	X	X	X	X	X	X		X		X	

Source: Author

4. Results

In order to analyze the different control algorithm strategies applied to solar tracking systems, Table 5 is presented, which classifies the literature found for each of the ties evaluated by percentage, applying a different strategy in the control algorithm. Table 5 summarizes the information recorded in Table 1 for open ties, in Table 2 for closed ties and finally, in Table 3 for hybrid ties, highlighting that 67.19% of the studies found in the literature applied an On- control. Off, being the most used, to its ease and low cost of implementation.

Additionally, of the analyzed research studies, 34.38% implemented an open cycle strategy, 45.31% applied a closed cycle and 20.31% used a hybrid cycle. It is important to mention that only works were considered in which the type of control algorithm used for the process was evidenced. However, the analysis of the information was not carried out under the same criteria, because both simulations and experimental processes were included, therefore, it cannot establish a universal criterion of the information. In addition, the focus of the analysis is directed to the fulfillment of the three types of control ties evaluated, a series of attributes applying the FCA methodology.

Table 5. Control algorithms for open-loop, closed-loop and hybrid-loop solar tracking techniques evaluated

Control algorithm strategy	Boucle			
	Open	Closed	Hybrid	100%
On-Off	20,31	32,81	14,06	67,19
PID	4,69	1,56	4,69	10,94
PI	4,69	0,00	0,00	4,69
Bayesian Network	0,00	0,00	1,56	1,56
Neuro Fuzzy Logic	0,00	1,56	0,00	1,56
FLC	1,56	3,13	0,00	4,69
ANFIS	1,56	0,00	0,00	1,56
Kinematic system	1,56	0,00	0,00	1,56
PID/FLC	0,00	3,13	0,00	3,13
LQR	0,00	1,56	0,00	1,56
SMC	0,00	1,56	0,00	1,56

Source: Author

On the other hand, Figure 4 represents the graphic diagram of the formal analysis of concepts carried out in the concept Explorer tool for the analysis of the control algorithms for solar tracking systems, based on the information consigned in the attribute matrix of **¡Error! No se encuentra el origen de la referencia.** This diagram relates open-loop, closed-loop and hybrid-loop control algorithms with the 11 proposed attributes.

The different nodes that are evidenced in Figure 4**¡Error! No se encuentra el origen de la referencia.** are identified as follows:

i. White-blue: contains only attributes
ii. Blue-black - Containers of attribute-related solar tracking technologies
iii. White-Black: Containers of only control algorithms
iv. White: Nodes that act as connectors between branches

With the above in mind, the analysis of the diagram is simple. The three technologies of solar tracking algorithms are related to attributes that, in turn, connect from top to bottom, to the node of origin. Attributes A1 "Single-axis Tracking System", A2 "Two-axis Tracking System", A3 "Classic Control Algorithm", A5 "Basic Control Unit" and A8 "Open Access Software Application" apply to the three types of control loop analyzed: "open", "closed" and "hybrid". From there, each control loop has another series of attributes that it shares with just one other loop or, failing that, it only applies to this one.

For its part, the Hybrid control loop additionally shares with the closed loop, compliance with attributes A4 "Modern control algorithm", A6 "Advanced control unit" and A11 "High application in photovoltaic systems". In turn, the closed control loop shares with the open control loop, compliance with attributes A7 "Availability of materials locally and A11" High application in photovoltaic systems. Finally, the open control loop also complies with attribute A9 "Low investment cost".

From this analysis, it is evident that hybrid control loop systems have high performance and precision, but that they require greater technical skill when building and assembling the system, directly incurring high investment costs. For its part, the open loop control algorithm presents flexibility in investment, construction, assembly and operation of the system, but it presents higher degrees of tracking error than the other two loops, directly affecting the efficiency of the system. Finally, the closed loop control algorithm presents acceptable efficiency values and its development is adapted to different types of software and hardware, presenting reasonable cost designs when compared with the good efficiency of the system, having low degrees of error of solar tracking.

IOP Conf. Series: Materials Science and Engineering 1253 (2022) 012017 doi:10.1088/1757-899X/1253/1/012017

Figure 4. FCA diagram of Information analysis

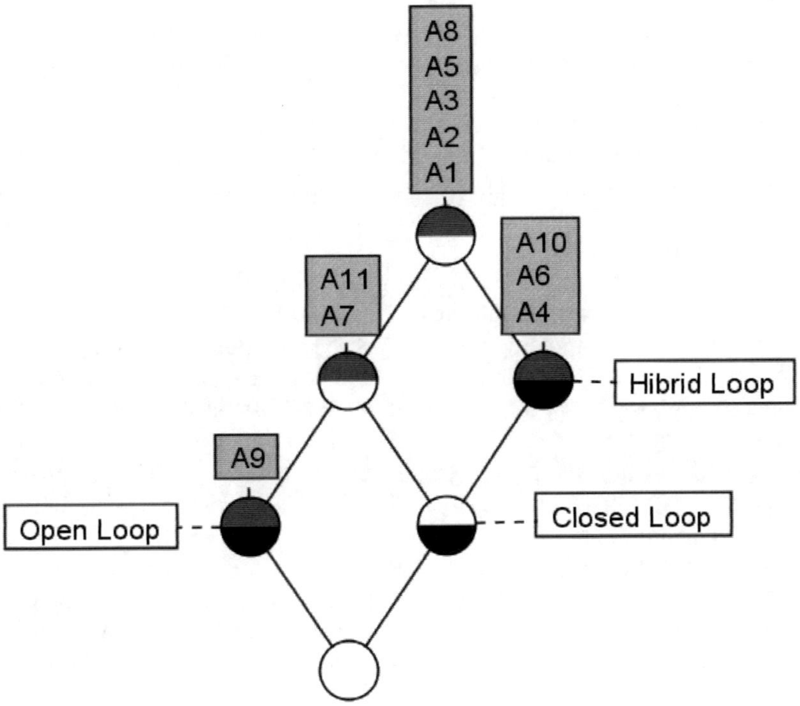

Source: Author

As an alternative to the proposed analysis approach, we can affirm that closed loop control algorithms are currently the most suitable to be developed in photovoltaic systems. However, single-loop algorithms when integrating GPS systems considerably improve their operability and may overcome the attributes that they do not comply with as the technique matures a little more. For their part, hybrid loop systems, despite their high reliability, still require greater technological developments in the research field and are projected as a promising alternative to be implemented in centralized and decentralized photovoltaic systems; It is important to mention that they are highly applied to solar concentration systems in centralized systems worldwide, but their high cost currently limits their integration with photovoltaic systems.

5. Conclusions

Solar tracking systems are an alternative to increase the efficiency of photovoltaic systems. In this article, a review was presented classifying into three types of control loops, open, closed and hybrid of strategies applied in solar trackers. The information was analyzed under the FCA methodology, evaluating characteristics such as flexibility in Hardware and Software, as well as the use of local resources for its development, offering sustainable solutions for its development and implementation, helping to overcome current problems for a specific problem. The most relevant conclusions of the work are presented below:

i. The majority of works in the literature do not present precision data or determination of the value of the solar tracking error.
ii. Of the works considered in this document,% use open circuit control strategies,% apply closed circuit strategy and% implement hybrid circuit control strategies. It can be concluded that the closed loop is the system with the greatest development today, offering tolerable

tracking errors and high efficiency improvements in the photovoltaic system. For its part, the open loop system, although it has applications, has high tracking errors and efficiencies below closed loop solar trackers. Finally, hybrid systems are projected as an alternative for the future, due to the low tracking error and the high efficiencies it offers, however, the cost of implementation is very high.

iii. The closed loop systems complied with 10 of the 11 attributes proposed for the FCA analysis, showing flexibility in their development, adapting to classic and modern control strategies, as well as the use of free or paid access software, as well as the use of low or high cost hardware, adapting to any need, that is, to simple or highly complex systems, with low tracking errors.

iv. The control unit for solar monitoring systems depends on the complexity of the implementation to be carried out, the capacity required by the system, directly incurring the autonomy of the equipment and its cost. The control units mainly used today are microcontrollers, due to their low cost and high capacity. As if, the use of computer equipment; applying software's like Matlab and LabVIEW.

v. There was a high interest by different authors in the development of control systems applied to solar trackers, however, there is no optimal system due to the differences that exist between all the designs, that is, each application is unique and its parameters satisfy specific needs of the installation.

Finally, it is important that the software development process and implementation of solar monitoring strategies continue, evaluating the different control loops in order to reach technological maturity in this type of systems that contribute significantly to improving the efficiency of the systems. of solar power generation. Similarly, technological development will have an important contribution to the growth of solar tracking systems, however, it is necessary to reduce costs in design, construction and implementation, as well as in maintenance, generating in turn, greater precision and accuracy. in solar tracking systems.

References

[1] T. H. Cetin, M. Kanoglu, y F. Bedir, «Integration of cryogenic energy storage and cryogenic organic cycle to geothermal power plants», *Geothermics*, vol. 87, p. 101830, sep. 2020, doi: 10.1016/j.geothermics.2020.101830.

[2] J. Stoklasa, P. Luukka, y M. Collan, «Possibilistic fuzzy pay–off method for real option valuation with application to research and development investment analysis», *Fuzzy Sets and Systems*, vol. 409, jul. 2020, doi: 10.1016/j.fss.2020.06.012.

[3] R. F. Fuentes-Morales *et al.*, «Control algorithms applied to active solar tracking systems: A review», *Solar Energy*, vol. 212, pp. 203-219, dic. 2020, doi: 10.1016/j.solener.2020.10.071.

[4] W. Nsengiyumva, S. G. Chen, L. Hu, y X. Chen, «Recent advancements and challenges in Solar Tracking Systems (STS): A review», *Renewable and Sustainable Energy Reviews*, vol. 81, pp. 250-279, ene. 2018, doi: 10.1016/j.rser.2017.06.085.

[5] R. Singh, S. Kumar, A. Gehlot, y R. Pachauri, «An imperative role of sun trackers in photovoltaic technology: A review», *Renewable and Sustainable Energy Reviews*, vol. 82, pp. 3263-3278, feb. 2018, doi: 10.1016/j.rser.2017.10.018.

[6] S. Skouri, A. Ben Haj Ali, S. Bouadila, M. Ben Salah, y S. Ben Nasrallah, «Design and construction of sun tracking systems for solar parabolic concentrator displacement», *Renewable and Sustainable Energy Reviews*, vol. 60, pp. 1419-1429, jul. 2016, doi: 10.1016/j.rser.2016.03.006.

[7] G. C. Lazaroiu, M. Longo, M. Roscia, y M. Pagano, «Comparative analysis of fixed and sun tracking low power PV systems considering energy consumption», *Energy Conversion and Management*, vol. 92, pp. 143-148, mar. 2015, doi: 10.1016/j.enconman.2014.12.046.

[8] F. A. Khalil, M. Asif, S. Anwar, S. ul Haq, y F. Illahi, «Solar Tracking Techniques and Implementation in Photovoltaic Power Plants: a Review: Solar Tracking Techniques and Implementation in Photovoltaic Power Plants», *Proceedings of the Pakistan Academy of Sciences: A. Physical and Computational Sciences*, vol. 54, n.º 3, Art. n.º 3, 2017.
[9] V. Sumathi, R. Jayapragash, A. Bakshi, y P. Kumar Akella, «Solar tracking methods to maximize PV system output – A review of the methods adopted in recent decade», *Renewable and Sustainable Energy Reviews*, vol. 74, pp. 130-138, jul. 2017, doi: 10.1016/j.rser.2017.02.013.
[10] S. Motahhir, A. El Hammoumi, y A. El Ghzizal, «The most used MPPT algorithms: Review and the suitable low-cost embedded board for each algorithm», *Journal of Cleaner Production*, vol. 246, p. 118983, feb. 2020, doi: 10.1016/j.jclepro.2019.118983.
[11] A. Z. Hafez, A. M. Yousef, y N. M. Harag, «Solar tracking systems: Technologies and trackers drive types – A review», *Renewable and Sustainable Energy Reviews*, vol. 91, pp. 754-782, ago. 2018, doi: 10.1016/j.rser.2018.03.094.
[12] N. AL-Rousan, N. A. M. Isa, y M. K. M. Desa, «Advances in solar photovoltaic tracking systems: A review», *Renewable and Sustainable Energy Reviews*, vol. 82, pp. 2548-2569, feb. 2018, doi: 10.1016/j.rser.2017.09.077.
[13] A. Awasthi *et al.*, «Review on sun tracking technology in solar PV system», *Energy Reports*, vol. 6, pp. 392-405, nov. 2020, doi: 10.1016/j.egyr.2020.02.004.
[14] IEC, «IEC 62817:2014 Photovoltaic systems - Design qualification of solar trackers». 2014. Accedido: jul. 27, 2021. [En línea]. Disponible en: https://www.une.org/encuentra-tu-norma/busca-tu-norma/iec/?c=7442
[15] W. Aipperspach, S. Bambrook, P. Trujillo, T. Zech, y F. R. Berenguel, «Evaluation of the IEC 62817 mechanical testing for the tracker validation», *AIP Conference Proceedings*, vol. 1679, n.º 1, p. 080001, sep. 2015, doi: 10.1063/1.4931542.
[16] L. Casajús y I. Muñoz, «Preliminary results after evaluating solar trackers based on IEC 62817:2014 Ed.1», *AIP Conference Proceedings*, vol. 1766, n.º 1, p. 110001, sep. 2016, doi: 10.1063/1.4962122.
[17] F. Sallaberry, R. Pujol-Nadal, M. Larcher, y M. H. Rittmann-Frank, «Direct tracking error characterization on a single-axis solar tracker», *Energy Conversion and Management*, vol. 105, pp. 1281-1290, nov. 2015, doi: 10.1016/j.enconman.2015.08.081.
[18] A. J. R. Del Rosario, A. T. Ubando, y A. B. Culaba, «Kinematic Synthesis and Analysis of a Watt II Six-Bar Linkage for the Design and Validation of an Open Loop Solar Tracking Mechanism», en *2019 IEEE 11th International Conference on Humanoid, Nanotechnology, Information Technology, Communication and Control, Environment, and Management (HNICEM)*, nov. 2019, pp. 1-6. doi: 10.1109/HNICEM48295.2019.9072698.
[19] Y. Zhu, J. Liu, y X. Yang, «Design and performance analysis of a solar tracking system with a novel single-axis tracking structure to maximize energy collection», *Applied Energy*, vol. 264, p. 114647, abr. 2020, doi: 10.1016/j.apenergy.2020.114647.
[20] K.-K. Chong *et al.*, «Integration of an On-Axis General Sun-Tracking Formula in the Algorithm of an Open-Loop Sun-Tracking System», *Sensors*, vol. 9, n.º 10, Art. n.º 10, oct. 2009, doi: 10.3390/s91007849.
[21] H. Fathabadi, «Comparative study between two novel sensorless and sensor based dual-axis solar trackers», *Solar Energy*, vol. 138, pp. 67-76, nov. 2016, doi: 10.1016/j.solener.2016.09.009.
[22] O. M. Pişirir y O. Bingöl, «Industrial PC Based Heliostat Control for Solar Power Towers», vol. 130, pp. 36-41, 2016.
[23] J. A. M. Medrano, *Fundamentos De Medición Y Control De Procesos*. Palibrio, 2017.
[24] I. Luque-Heredia, G. Quéméré, R. Cervantes, O. Laurent, E. Chiappori, y J. Y. Chong, «The Sun Tracker in Concentrator Photovoltaics», en *Next Generation of Photovoltaics: New Concepts*, A. B. Cristóbal López, A. Martí Vega, y A. Luque López, Eds. Berlin, Heidelberg: Springer, 2012, pp. 61-93. doi: 10.1007/978-3-642-23369-2_3.

[25] N. AL-Rousan, N. A. Mat Isa, y M. K. Mat Desa, «Efficient single and dual axis solar tracking system controllers based on adaptive neural fuzzy inference system», *Journal of King Saud University - Engineering Sciences*, vol. 32, n.º 7, pp. 459-469, nov. 2020, doi: 10.1016/j.jksues.2020.04.004.
[26] M. Debbache *et al.*, «Mathematical Model of an Azimuthal- Elevation Tracking System of Small Scale Heliostat», vol. 38, n.º 9, pp. 111-120, 2016.
[27] H. Fathabadi, «Novel high efficient offline sensorless dual-axis solar tracker for using in photovoltaic systems and solar concentrators», *Renewable Energy*, vol. 95, pp. 485-494, sep. 2016, doi: 10.1016/j.renene.2016.04.063.
[28] M. Ikhwan, Mardlijah, y C. Imron, «Model predictive control on dual axis solar tracker using Matlab/Simulink simulation», en *2018 International Conference on Information and Communications Technology (ICOIACT)*, mar. 2018, pp. 784-788. doi: 10.1109/ICOIACT.2018.8350791.
[29] C.-K. Yang, T.-C. Cheng, C.-H. Cheng, C.-C. Wang, y C.-C. Lee, «Open-loop altitude-azimuth concentrated solar tracking system for solar-thermal applications», *Solar Energy*, vol. 147, pp. 52-60, may 2017, doi: 10.1016/j.solener.2017.03.014.
[30] M. H. M. Sidek, N. Azis, W. Z. W. Hasan, M. Z. A. Ab Kadir, S. Shafie, y M. A. M. Radzi, «Automated positioning dual-axis solar tracking system with precision elevation and azimuth angle control», *Energy*, vol. 124, pp. 160-170, abr. 2017, doi: 10.1016/j.energy.2017.02.001.
[31] Z. Mi, J. Chen, N. Chen, Y. Bai, R. Fu, y H. Liu, «Open-loop solar tracking strategy for high concentrating photovoltaic systems using variable tracking frequency», *Energy Conversion and Management*, vol. 117, pp. 142-149, jun. 2016, doi: 10.1016/j.enconman.2016.03.009.
[32] J. C. M. Carvalho, D. Martins, R. Simoni, y H. Simas, *Multibody Mechatronic Systems: Proceedings of the MUSME Conference held in Florianópolis, Brazil, October 24-28, 2017*. Springer, 2017.
[33] F. Bedaouche, A. Gama, A. Hassam, R. Khelifi, y M. Boubezoula, «Fuzzy PID Control of a DC Motor to Drive a Heliostat», en *2017 International Renewable and Sustainable Energy Conference (IRSEC)*, dic. 2017, pp. 1-6. doi: 10.1109/IRSEC.2017.8477396.
[34] R. C. Allil *et al.*, «Solar tracker development based on a POF bundle and Fresnel lens applied to environment illumination and microalgae cultivation», *Solar Energy*, vol. 174, pp. 648-659, nov. 2018, doi: 10.1016/j.solener.2018.09.061.
[35] R. C. 2 Mohamed Redha Rezoug *1, «The Optimal Angles Of A Dual-Axis Tracking System By Pre-Programmed Method Using A Microcontroller», mar. 2018, doi: 10.5281/ZENODO.1207029.
[36] J. Ruelas, F. Muñoz, B. Lucero, y J. Palomares, «PV Tracking Design Methodology Based on an Orientation Efficiency Chart», *Applied Sciences*, vol. 9, n.º 5, Art. n.º 5, ene. 2019, doi: 10.3390/app9050894.
[37] M. E. H. Chowdhury, A. Khandakar, B. Hossain, y R. Abouhasera, «A Low-Cost Closed-Loop Solar Tracking System Based on the Sun Position Algorithm», *Journal of Sensors*, vol. 2019, p. e3681031, feb. 2019, doi: 10.1155/2019/3681031.
[38] J. Canada-Bago, J.-A. Fernandez-Prieto, M.-A. Gadeo-Martos, y P. Perez-Higueras, «Knowledge-Based Sensors for Controlling A High-Concentration Photovoltaic Tracker», *Sensors*, vol. 20, n.º 5, Art. n.º 5, ene. 2020, doi: 10.3390/s20051315.
[39] L. F. Saavedra Quintero y D. F. Sánchez Martínez, «Automatización de un sistema de seguimiento solar en un eje para un prototipo de colector lineal tipo Fresnel de las Unidades Tecnológicas de Santander», abr. 2021, Accedido: sep. 09, 2021. [En línea]. Disponible en: http://repositorio.uts.edu.co:8080/xmlui/handle/123456789/5828
[40] A. Saymbetov *et al.*, «Dual-axis schedule tracker with an adaptive algorithm for a strong scattering of sunbeam», *Solar Energy*, vol. 224, pp. 285-297, ago. 2021, doi: 10.1016/j.solener.2021.06.024.

[41] M. Abdollahpour, M. R. Golzarian, A. Rohani, y H. Abootorabi Zarchi, «Development of a machine vision dual-axis solar tracking system», *Solar Energy*, vol. 169, pp. 136-143, jul. 2018, doi: 10.1016/j.solener.2018.03.059.
[42] G. E. G. Mustafa, B. A. M. Sidahmed, y M. O. Nawari, «The Improvement of LDR Based Solar Tracker's Action using Machine Learning», en *2019 IEEE Conference on Energy Conversion (CENCON)*, oct. 2019, pp. 230-235. doi: 10.1109/CENCON47160.2019.8974834.
[43] L. Salgado-Conrado, «A review on sun position sensors used in solar applications», *Renewable and Sustainable Energy Reviews*, vol. 82, pp. 2128-2146, feb. 2018, doi: 10.1016/j.rser.2017.08.040.
[44] M. G. Satué, F. Castaño, M. G. Ortega, y F. R. Rubio, «Auto-calibration method for high concentration sun trackers», *Solar Energy*, vol. 198, pp. 311-323, mar. 2020, doi: 10.1016/j.solener.2019.12.073.
[45] D. Mazumdar, D. Sinha, S. Panja, y D. K. Dhak, «Design of LQR controller for solar tracking system», en *2015 IEEE International Conference on Electrical, Computer and Communication Technologies (ICECCT)*, mar. 2015, pp. 1-5. doi: 10.1109/ICECCT.2015.7226000.
[46] A. Díaz, S. Keshtkar, J. A. Moreno, y E. Hernandez, «Design and Control Strategy of a Low-Cost Parallel Robot for Precise Solar Tracking», en *Advances in Service and Industrial Robotics*, Cham, 2019, pp. 342-350. doi: 10.1007/978-3-030-00232-9_36.
[47] R. Garrido y A. Díaz, «Cascade closed-loop control of solar trackers applied to HCPV systems», *Renewable Energy*, vol. 97, pp. 689-696, nov. 2016, doi: 10.1016/j.renene.2016.06.022.
[48] J. A. Carballo, J. Bonilla, L. Roca, y M. Berenguel, «New low-cost solar tracking system based on open source hardware for educational purposes», *Solar Energy*, vol. 174, pp. 826-836, nov. 2018, doi: 10.1016/j.solener.2018.09.064.
[49] C. Jamroen, P. Komkum, S. Kohsri, W. Himananto, S. Panupintu, y S. Unkat, «A low-cost dual-axis solar tracking system based on digital logic design: Design and implementation», *Sustainable Energy Technologies and Assessments*, vol. 37, p. 100618, feb. 2020, doi: 10.1016/j.seta.2019.100618.
[50] C. Morón, D. Ferrández, P. Saiz, G. Vega, y J. P. Díaz, «New Prototype of Photovoltaic Solar Tracker Based on Arduino», *Energies*, vol. 10, n.º 9, Art. n.º 9, sep. 2017, doi: 10.3390/en10091298.
[51] A. E. Hammoumi, S. Motahhir, A. E. Ghzizal, A. Chalh, y A. Derouich, «A simple and low-cost active dual-axis solar tracker», *Energy Science & Engineering*, vol. 6, n.º 5, pp. 607-620, 2018, doi: 10.1002/ese3.236.
[52] S. Yilmaz, H. Riza Ozcalik, O. Dogmus, F. Dincer, O. Akgol, y M. Karaaslan, «Design of two axes sun tracking controller with analytically solar radiation calculations», *Renewable and Sustainable Energy Reviews*, vol. 43, pp. 997-1005, mar. 2015, doi: 10.1016/j.rser.2014.11.090.
[53] Z. El Kadmiri, O. El Kadmiri, L. Masmoudi, y M. N. Bargach, «A Novel Solar Tracker Based on Omnidirectional Computer Vision», *Journal of Solar Energy*, vol. 2015, pp. 1-6, ene. 2015, doi: 10.1155/2015/149852.
[54] C.-C. Wei, Y.-C. Song, C.-C. Chang, y C.-B. Lin, «Design of a Solar Tracking System Using the Brightest Region in the Sky Image Sensor», *Sensors*, vol. 16, n.º 12, Art. n.º 12, dic. 2016, doi: 10.3390/s16121995.
[55] E. Normanyo y A. Akparibo, «A Solar Radiation Tracker for Solar Energy Optimisation», *British Journal of Applied Science & Technology*, vol. 14, pp. 1-12, ene. 2016, doi: 10.9734/BJAST/2016/22955.
[56] M. M. Sabir y T. Ali, «Optimal PID controller design through swarm intelligence algorithms for sun tracking system», *Applied Mathematics and Computation*, vol. 274, pp. 690-699, feb. 2016, doi: 10.1016/j.amc.2015.11.036.
[57] S. Keshtkar, J. Keshtkar, y A. Poznyak, «Adaptive sliding mode control for solar tracker orientation», en *2016 American Control Conference (ACC)*, jul. 2016, pp. 6543-6548. doi: 10.1109/ACC.2016.7526700.

[58] E. Kiyak y G. Gol, «A comparison of fuzzy logic and PID controller for a single-axis solar tracking system», *Renewables*, vol. 3, n.º 1, p. 7, feb. 2016, doi: 10.1186/s40807-016-0023-7.
[59] C.-H. Huang, H.-Y. Pan, y K.-C. Lin, «Development of Intelligent Fuzzy Controller for a Two-Axis Solar Tracking System», *Applied Sciences*, vol. 6, n.º 5, Art. n.º 5, may 2016, doi: 10.3390/app6050130.
[60] S. Makhija, A. Khatwani, Mohd. F. Khan, V. Goel, y M. M. Roja, «Design amp; implementation of an automated dual-axis solar tracker with data-logging», en *2017 International Conference on Inventive Systems and Control (ICISC)*, ene. 2017, pp. 1-4. doi: 10.1109/ICISC.2017.8068708.
[61] S. Hussain, A. I. Akbar, M. W. Siddiq, y Aziz, «Microcontroller Based Dual Axis Sun Tracking System for Maximum Solar Energy Generation», vol. 5, n.º 1, pp. 23-27, 2017.
[62] D. A. Flores-Hernández, S. Palomino-Resendiz, N. Lozada-Castillo, A. Luviano-Juárez, y I. Chairez, «Mechatronic design and implementation of a two axes sun tracking photovoltaic system driven by a robotic sensor», *Mechatronics*, vol. 47, pp. 148-159, nov. 2017, doi: 10.1016/j.mechatronics.2017.09.014.
[63] F. M. Hoffmann, R. F. Molz, J. V. Kothe, E. O. B. Nara, y L. P. C. Tedesco, «Monthly profile analysis based on a two-axis solar tracker proposal for photovoltaic panels», *Renewable Energy*, vol. 115, pp. 750-759, ene. 2018, doi: 10.1016/j.renene.2017.08.079.
[64] M. Ghassoul, «Single axis automatic tracking system based on PILOT scheme to control the solar panel to optimize solar energy extraction», *Energy Reports*, vol. 4, pp. 520-527, nov. 2018, doi: 10.1016/j.egyr.2018.07.001.
[65] Y. Su, Y. Zhang, y L. Shu, «Experimental study of using phase change material cooling in a solar tracking concentrated photovoltaic-thermal system», *Solar Energy*, vol. 159, pp. 777-785, ene. 2018, doi: 10.1016/j.solener.2017.11.045.
[66] J. A. Carballo, J. Bonilla, M. Berenguel, J. Fernández-Reche, y G. García, «New approach for solar tracking systems based on computer vision, low cost hardware and deep learning», *Renewable Energy*, vol. 133, pp. 1158-1166, abr. 2019, doi: 10.1016/j.renene.2018.08.101.
[67] A. A. Okandeji, M. B. Olajide, G. O. Olasunkanmi, y Z. O. Jagun, «Analysis and implementation of a solar tracking rack system», *Nigerian Journal of Technology*, vol. 39, n.º 3, Art. n.º 3, sep. 2020, doi: 10.4314/njt.v39i3.29.
[68] R. N. Hasanah, S. Aditya Bagus, M. Eka, N. Tri, y T. Taufik, «Computer-based solar tracking system for PV energy yield improvement», *International Journal of Power Electronics and Drive Systems; Yogyakarta*, vol. 11, n.º 2, pp. 743-751, 2020, doi: 10.11591/ijpeds.v11.i2.pp743-751.
[69] J. J. Ontiveros, C. D. Ávalos, F. Loza, N. D. Galán, y G. J. Rubio, «Evaluation and Design of Power Controller of Two-Axis Solar Tracking by PID and FL for a Photovoltaic Module», *International Journal of Photoenergy*, vol. 2020, p. e8813732, jul. 2020, doi: 10.1155/2020/8813732.
[70] J. A. Forero Monsalve y B. A. Jaimes Grimaldos, «Desarrollo de un sistema automatizado de seguimiento del sol para dar movimiento a un colector solar cilindro parabólico a través de la implementación de un chip Esp32 y un Servomotor monitoreados por un miniordenador Raspberry», ene. 2021, Accedido: sep. 09, 2021. [En línea]. Disponible en: http://repositorio.uts.edu.co:8080/xmlui/handle/123456789/5182
[71] C. Jamroen, C. Fongkerd, W. Krongpha, P. Komkum, A. Pirayawaraporn, y N. Chindakham, «A novel UV sensor-based dual-axis solar tracking system: Implementation and performance analysis», *Applied Energy*, vol. 299, p. 117295, oct. 2021, doi: 10.1016/j.apenergy.2021.117295.
[72] P. Munanga, S. Chinguwa, W. R. Nyemba, y C. Mbohwa, «Design for manufacture and assembly of an intelligent single axis solar tracking system», *Procedia CIRP*, vol. 91, pp. 571-576, ene. 2020, doi: 10.1016/j.procir.2020.03.109.
[73] Y. M. Safan, S. Shaaban, y M. I. A. El-Sebah, «Hybrid control of a solar tracking system using SUI-PID controller», en *2017 Sensors Networks Smart and Emerging Technologies (SENSET)*, sep. 2017, pp. 1-4. doi: 10.1109/SENSET.2017.8125035.

[74] H.-Y. Yeh y C.-D. Lee, «The Logic-Based Supervisor Control for Sun-Tracking System of 1 MW HCPV Demo Plant: Study Case», *Applied Sciences*, vol. 2, n.º 1, Art. n.º 1, mar. 2012, doi: 10.3390/app2010100.
[75] G. Liu, A. O. Baba, y L. Zhu, «Hybrid controller for dual axes solar tracking system», en *2017 36th Chinese Control Conference (CCC)*, jul. 2017, pp. 3203-3207. doi: 10.23919/ChiCC.2017.8027850.
[76] H. Kang, T. Hong, y M. Lee, «Technical performance analysis of the smart solar photovoltaic blinds based on the solar tracking methods considering the climate factors», *Energy and Buildings*, vol. 190, pp. 34-48, may 2019, doi: 10.1016/j.enbuild.2019.02.013.
[77] S. Seme, B. Štumberger, y M. Hadžiselimović, «A novel prediction algorithm for solar angles using second derivative of the energy for photovoltaic sun tracking purposes», *Solar Energy*, vol. 137, pp. 201-211, nov. 2016, doi: 10.1016/j.solener.2016.08.001.
[78] J. Zhang, Z. Yin, y P. Jin, «Error analysis and auto correction of hybrid solar tracking system using photo sensors and orientation algorithm», *Energy*, vol. 182, pp. 585-593, sep. 2019, doi: 10.1016/j.energy.2019.06.032.
[79] H. Kang, T. Hong, y M. Lee, «A new approach for developing a hybrid sun-tracking method of the intelligent photovoltaic blinds considering the weather condition using data mining technique», *Energy and Buildings*, vol. 209, p. 109708, feb. 2020, doi: 10.1016/j.enbuild.2019.109708.
[80] K. Jianwattananukul, K. Locharoenrat, y S. Lekchaum, «Design and Construction of New Hybrid Solar Tracking System», *Applied Mechanics and Materials*, vol. 851, pp. 510-515, 2016, doi: 10.4028/www.scientific.net/AMM.851.510.
[81] K. Mao, F. Liu, y I. R. Ji, «Design of ARM-Based Solar Tracking System», en *2018 37th Chinese Control Conference (CCC)*, jul. 2018, pp. 7394-7398. doi: 10.23919/ChiCC.2018.8483773.
[82] Q. Xie, Y. Xiao, X. Wang, D. Liu, y Z. Shen, «Heliostat Cluster Control for the Solar Tower Power Plant Based on Leader-Follower Strategy», *IEEE Access*, vol. 7, pp. 135031-135039, 2019, doi: 10.1109/ACCESS.2019.2940618.
[83] K.-H. Kim y S.-B. Cho, «An efficient concentrative photovoltaic solar system with Bayesian selection of optimal solar tracking algorithms», *Applied Soft Computing*, vol. 83, p. 105618, oct. 2019, doi: 10.1016/j.asoc.2019.105618.
[84] B. Alorda, R. Pujol-Nadal, G. Rodriguez-Navas, A. Moià-Pol, y V. Martínez-Moll, «Collaborative distributed sun-tracking control system for building integration with minimal plant area and maximum energy-conversion efficiency», *International Journal of Electrical Power & Energy Systems*, vol. 70, pp. 52-60, sep. 2015, doi: 10.1016/j.ijepes.2015.01.038.
[85] M. Burhan, S. J. Oh, K. J. E. Chua, y K. C. Ng, «Double lens collimator solar feedback sensor and master slave configuration: Development of compact and low cost two axis solar tracking system for CPV applications», *Solar Energy*, vol. 137, pp. 352-363, nov. 2016, doi: 10.1016/j.solener.2016.08.035.
[86] G. B. EzhilJenekkha y M. MarsalineBeno, «Comparative analysis of open loop and non-isolated control for solar panel», *Materials Today: Proceedings*, jun. 2021, doi: 10.1016/j.matpr.2021.05.635.
[87] R. Ahmed *et al.*, «Computer vision and photosensor based hybrid control strategy for a two-axis solar tracker - Daylighting application», *Solar Energy*, vol. 224, pp. 175-183, ago. 2021, doi: 10.1016/j.solener.2021.05.077.

AUTHOR INDEX